T0076748

Advances in Industrial Control

Series editors

For further volumes:
http://www.springer.com/series/1412

Yongseob Lim · Ravinder Venugopal
A Galip Ulsoy

Process Control for Sheet-Metal Stamping

Process Modeling, Controller Design and Shop-Floor Implementation

Yongseob Lim
Samsung Techwin R&D Center
Seongnam-si
Korea, Republic of South Korea

Ravinder Venugopal
Intellicass Inc.
Montreal, QC
Canada

A Galip Ulsoy
Department of Mechanical Engineering
University of Michigan
Ann Arbor, MI
USA

ISSN 1430-9491 ISSN 2193-1577 (electronic)
ISBN 978-1-4471-6283-4 ISBN 978-1-4471-6284-1 (eBook)
DOI 10.1007/978-1-4471-6284-1
Springer London Heidelberg New York Dordrecht

Library of Congress Control Number: 2013955567

Printed on acid-free paper

Springer is part of Springer Science+Business Media (www.springer.com)

Foreword

The series *Advances in Industrial Control* aims to report and encourage technology transfer in control engineering. The rapid development of control technology has an impact on all areas of the control discipline. New theory, new controllers, actuators, sensors, new industrial processes, computer methods, new applications, new philosophies..., new challenges. Much of this development work resides in industrial reports, feasibility study papers and the reports of advanced collaborative projects. The series offers an opportunity for researchers to present an extended exposition of such new work in all aspects of industrial control for wider and rapid dissemination.

The iron and steel industry along with other metal-producing industries (aluminium, for example) contain a wealth of control systems and control system technologies that make them endlessly fascinating. Control engineers working in these fields will undoubtedly spend as much time understanding the constituent processes as they will in designing and implementing actual control systems. This is a fairly common experience for the industrial control engineer that is not often mentioned in university undergraduate control courses.

Consider the production of steel sheet that finds its way into automobile bodies, white goods and a myriad of other consumer goods. The production process starts in the casting shop, followed by the reheat furnace, then the hot rolling mill followed by the cold rolling mill. And if the sheet is to be tinned or coated then the sheet undergoes a trip to a coating line to complete the process. At every step of the way, it is the control-engineering input that ensures product quality is maintained and often improved and that the process remains safe and secure for the operating workforce. For this particular branch of the metals industry, the *Advances in Industrial Control* monograph series has two entries that deal with sheet processes:

- *Identification and Control of Sheet and Film Processes* by Andrew P. Featherstone, Jeremy G. VanAntwerp and Richard D. Braatz (ISBN 978-1-85233-305-8, 2000) and
- *Tandem Cold Metal Rolling Mill Control* by John Pittner and Marwan A. Simaan (ISBN 978-1-85729-066-3, 2011)

Taking a more general approach to industrial automation and performance monitoring but demonstrating its methods with an extended application also taken from sheet-production processes is the monograph:

- *Control Performance Management in Industrial Automation* by Mohieddine Jelali (ISBN 978-1-4471-4545-5, 2012).

It is therefore of considerable value to the *Advances in Industrial Control* series to be able to add to its list the monograph *Process Control for Sheet-Metal Stamping* by Yongseob Lim, Ravinder Venugopal and A. Galip Ulsoy that describes the processes, the control systems and the technology that uses the *output* of these sheet-production lines. In this monograph the reader will find a focused introduction to sheet stamping processes and their related control technologies; this comprises an introductory chapter, three further stamping technology chapters and a chapter on a laboratory development for stamping process control investigations. The next three chapters report control design, simulations and experimental validations of the control strategies proposed. Interestingly, the control techniques used are SIMO PI controllers, the relay experiment for PI tuning and then model reference adaptive control (MRAC) strategies where adaptation is used to improve product quality and overcome the uncertainties and disturbances in the process. As the control strategies are developed, care is taken to note the implications and costs of controller implementation issues, and the economic benefits that accrue from using more advanced methods. Developing controls that can be translated to the industrial production line and the shop floor is an overriding objective of the work reported. In the concluding chapter, the point is made that although difficult to cost economically, the introduction of the new control system has made the metal stamping process safer and cleaner for the shop-floor workers and operating staff.

Some readers will value the monograph as a repository of state-of-the-art information about the technology and control of stamping processes. Other readers will find the monograph an exemplary case study in developing and validating new and advanced control solutions for a widespread industrial process that is ready for technology advancement. This monograph is a very worthy addition to the *Advances in Industrial Control* series.

Glasgow, Scotland, UK

M. J. Grimble
M. A. Johnson

Preface

Stamping is a long-established, widely used industrial process for economical high-volume production. It is used extensively in the automotive industry, as well as for production of white goods and many other products. In this book we present an approach, based on process control, to improve stamped part quality at reduced cost by eliminating tearing, wrinkling and springback. The concept is straightforward: measure punch forces and then adjust the blank holder (i.e., binder) forces (i.e., how tightly we hold the blank material in place) at various locations around its periphery and at various times during the stamping process to properly control the draw-in of blank material into the die. Of course, how to do this is the challenge! This book describes in detail how this simple goal can be achieved through real-time control technology.

A reconfigurable set of hydraulic actuators (e.g., 12–24) is placed under the die to enable the control of the blank holder forces at various locations around the die periphery. These blank holder forces at each actuator are varied during the short duration (e.g., <1 sec) of the press stroke. The careful design of a controller, termed the *machine controller*, is needed to ensure that the desired blank holder forces are achieved at each hydraulic actuator and at each instant in time during the press stroke. Furthermore, we also measure the punch force during stamping, and design another controller, termed the *process controller*, to ensure that the desired punch force values are achieved during stamping despite the presence of disturbances (e.g., lubrication or material thickness variations). Maintaining the desired punch force leads to consistent draw-in of blank material and improves stamped part quality by eliminating wrinkling, tearing and springback.

In this book we describe the methods for designing these controllers, and present experimental validation results from die try-out tests. The proposed system has also been evaluated in pilot tests in production and has also been shown to improve the formability of hard-to-form materials, such as lightweight alloys.

This book is the result of a multi-year research collaboration among the authors. We would like to thank the State of Michigan's Twenty-first Century Fund for their financial support of this research project, and also thank our industrial collaborators Troy Design And Manufacturing (TDM) Company, Ogihara America Corporation and OPAL-RT Technologies.

The real-time computer control equipment used was provided by OPAL-RT. The die try-out tests, and the experimental results presented in the book, were based on extensive studies carried out by Dr. Yongseob Lim and Dr. Ravi Venugopal at TDM in Warren, Michigan with considerable assistance from TDM management, engineers and operators. The stamping process control system described in this book was also evaluated by Dr. Lim and Dr. Venugopal in pilot production tests at Ogihara's plant in Howell, Michigan. The research work provided the basis for the doctoral dissertation of Dr. Yongseob Lim, under the supervision of Professor A. Galip Ulsoy, at the University of Michigan, Ann Arbor. The company Intellicass, Inc. (see http://www.intellicass.com/) was established by Dr. Venugopal and utilizes the research described in this book.

We hope that this book will provide a foundation for the widespread use of process control systems in stamping, and thereby provide the significant benefits to both producers and consumers that we have described in Chap. 9.

August 31, 2013

Yongseob Lim
Ravinder Venugopal
A Galip Ulsoy

Contents

Chapter 1
Introduction

Abstract Controlling a manufacturing process can increase productivity, reduce cost and improve quality. In this chapter the role of automation and process control in manufacturing is introduced. In a manufacturing process control system appropriate indicators of process performance are measured, and then used to adjust the process to achieve consistency in the presence of disturbances. Process control has been developed for many manufacturing processes, such as machining and semiconductor fabrication. Process control for stamping is a relatively new development and is introduced here and then discussed in detail in the remainder of this book.

1.1 Manufacturing Automation

Manufacturing, together with agriculture and mining, is one of the basic mechanisms of wealth creation in society. It is essential for sustaining the service industries (e.g., engineering, finance, retail, education, entertainment) in any healthy economy (Cohen and Zysman 1987). Consequently, manufacturing matters, and it is critical for economic development. The overarching goals in any manufacturing operation are to increase productivity, reduce cost, and to improve quality. To achieve these goals, and to be competitive in manufacturing, both operations engineering and manufacturing automation are known to play critical and complementary roles (Womack et al. 1990). Falling behind in manufacturing innovation, and manufacturing automation, can lead to major economic consequences: "A sustained weakness in manufacturing capabilities could endanger the technology base of the country" (Cohen and Zysman 1988). Manufacturing systems, to be competitive, must also be responsive to the changing global marketplace, and automation can also play a major role in providing such flexibility and reconfigurability (Koren et al. 1999; Koren 2010).

Computer controlled machine tools were introduced about 50 years ago, and have had a major impact on industrial production (Koren 1983). Numerically controlled (NC) machine tools were developed by the Parson's Machine Tool

Y. Lim et al., *Process Control for Sheet-Metal Stamping*,
Advances in Industrial Control, DOI: 10.1007/978-1-4471-6284-1_1,

Company in Michigan and the Servomechanisms Laboratory at MIT in the 1950s. These reprogrammable but hard-wired digital devices represented the state-of-the-art into the 1960s. During the 1960s and 1970s, computers became not only more powerful, but also less expensive and more reliable. The servo-control function (including multi-axis interpolators) became implemented using on-board computers rather than hard-wired digital circuits. These so-called computer numerically controlled (CNC) systems, because of their powerful computing capabilities, led to advances in the interpolators and in the servo control loops.

Typically CNC machines ensure the correct positioning and movement of a tool relative to a workpiece, and are used in milling, drilling, turning, grinding, inspection, welding, semiconductor fabrication, and in many other manufacturing processes. Today they are the workhorses of any major manufacturing operation.

1.2 Process Control in Manufacturing

The availability of significant on-board computing power on CNC machines facilitated the introduction of an additional level of control, i.e. process control (sometimes called "adaptive control" in the manufacturing literature), which can be used to improve process performance in the presence of disturbances. For example, increase metal removal rates, improve part quality and/or prevent process failures (Ulsoy et al. 1983; Ulsoy and Koren 1993; Rashap et al. 1995). As shown in Fig. 1.1 this is a process-level feedback controller that measures some appropriate indicator of process performance (e.g., cutting force) and compares it to a reference value (typically determined by off-line process optimization) and then makes adjustments to the process via a CNC machine. The CNC machine itself typically contains additional feedback loops (e.g., servo loops to control position and speed). Thus, those faster responding machine control feedback loops can be viewed as being nested inside the process control feedback loop. The process controller assumes the availability of a lower-level machine controller, and builds on its capabilities to provide consistent performance in the presence of disturbance inputs.

As an example of these process controllers for a manufacturing process consider a model reference adaptive force controller for a milling process (Lauderbaugh and Ulsoy 1989). In a typical slot milling operation, the CNC machine controller will position the tool (i.e., milling cutter) relative to the workpiece based on a user supplied part program. The part program also specifies the spindle speed at which the tool is rotated, as well as the feedrate (i.e., velocity) at which the rotating tool is moved through the workpiece material at a constant desired depth-of-cut. These tasks are all performed by the CNC machine controller. If the workpiece has an irregular (e.g., tapered, stepped or rough casting) geometry the actual depth-of-cut will change during the operation, and act as a process disturbance. This can lead to large cutting forces and possible tool failure. If the programmed depth-of-cut is conservatively reduced to eliminate such problems, then

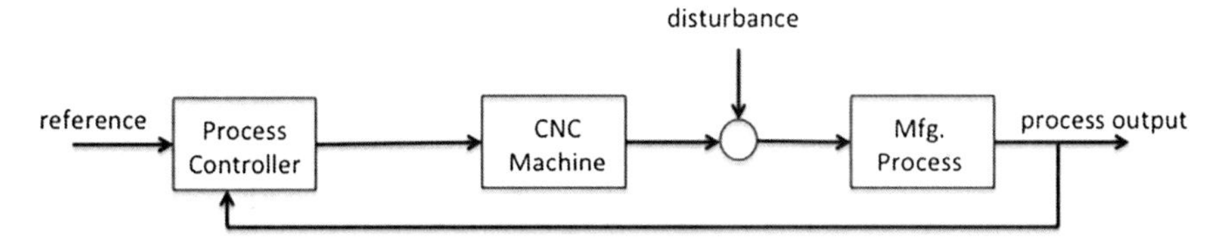

Fig. 1.1 Process controller for a manufacturing process

the metal removal rate is reduced and productivity suffers. Consequently, one can add a process control level (as in Fig. 1.1) to supplement the CNC machine control. At the process control level the cutting force (e.g., average or maximum resultant force) is measured and then fed back to compare to a desired reference force, which has been selected to ensure fast and safe operation for that tool. The process controller then adjusts the feedrate of the tool to achieve safe operation with high productivity.

A discrete-time transfer function model of a two-axis slot milling process can be written as (Lauderbaugh and Ulsoy 1988),

$$\frac{F_R(z)}{V(z)} = \frac{b_0 z + b_1}{z^2 + a_1 z + a_2} \tag{1.1}$$

where V is a voltage signal proportional to the machine feedrate, and F_R is the measured resultant force. The parameters depend upon the spindle speed, feed, depth-of-cut, workpiece material properties, etc. Thus, they must be estimated on-line for effective control of the resultant force based upon manipulation of the feedrate (i.e., V). The goal is to maintain the resultant force at the reference level, R, which is selected to maintain high metal removal rates without problems of tool breakage. The process zero can vary, but was experimentally found to be inside the unit circle for all cutting situations at the 50 ms sampling period that was used (Lauderbaugh and Ulsoy 1988). Therefore, direct adaptive control methods can be employed. A model reference adaptive controller (MRAC) was designed and evaluated in laboratory tests for two different workpiece materials and for changing depths of cut (Lauderbaugh and Ulsoy 1989). The MRAC design gave satisfactory performance despite the process variations. Experimental results are shown in Fig. 1.2 for machining of a 1020 CR steel workpiece with step changes in depth-of-cut from 2 to 3 to 4 mm. The desired value of $R = 400$ N is maintained despite these large changes in depth-of-cut at a spindle-speed of 550 rev/min. Note that initially the feedrate saturates, and the reference value of 400 N cannot be achieved at the 2 mm depth-of-cut. Also, initial transients occur due to the parameter adaptation. This controller also gave good results when used in machining other materials (e.g., aluminum) without any additional controller tuning (Lauderbaugh and Ulsoy 1989).

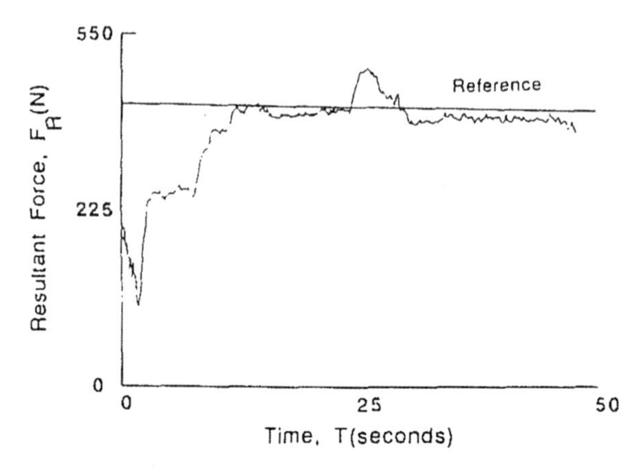

Fig. 1.2 Resultant force versus time with MRAC in slot milling of a steel workpiece with step changes in depth-of-cut (Lauderbaugh and Ulsoy 1989)

1.3 Stamping Process

Sheet metal parts are all around us: beverage cans, metal desks, white goods, car bodies, aircraft fuselages, etc. They are lightweight, strong, and can take on complex shapes. The sheet metal stamping process is characterized by very high production rates, low labor costs, but high equipment and tooling costs. Thus, this process is ideally suited for high-volume production.

The sheet metal blanks used in stamping are typically made of low-carbon steel, because of its low cost, good strength and excellent formability. The formability of various sheet metals is typically determined by marking the sheet with a grid of small circles, and then stretching it over a punch (see Fig. 1.3). The deformation of the circles is measured in regions where tearing has occurred and used to construct a forming-limit diagram (Kalpakjian and Schmid 2001; Marciniak et al. 2002; Hosford and Caddell 2011).

In applications where lightweight is important, aluminum, or alloys of steel with magnesium and titanium are also used. These lighter-weight materials are typically more expensive and less ductile and harder to form. Thus, developing tooling to produce defect-free parts using these lightweight materials is difficult.

As shown in Fig. 1.4, a typical stamping press consists of a punch (upper die), a die, and blank holder (binder), which holds the sheet metal in place during the punch stroke (i.e., while the punch is lowered into the die). The stroke (e.g., 50 mm) depends on the desired part geometry. The die design will often include appropriately placed drawbeads to help regulate the material flow into the die. Commercial stamping operations are typically done at high pressure (e.g., 10 kPa) and high speed, thus, leading to short duration (e.g., 0.5 s). The sheet metal material is plastically deformed, and flows into the die cavity and conforms to its shape. Proper design of the die allows complex shapes to be produced rapidly and cost effectively. Blank holder force (F_b) and punch force (F_p) must be properly selected to hold the sheet metal blank in place, through friction forces, while still allowing the sheet metal to plastically deform and flow into the die cavity.

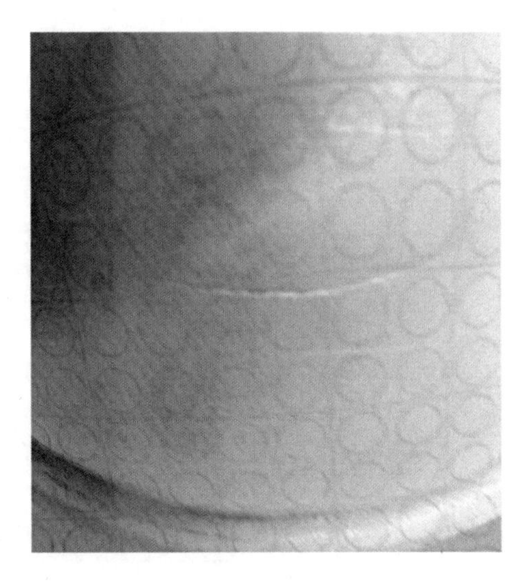

Fig. 1.3 Grid of circles for formability testing with a spherical punch

Typically, die design, and selection of nominal process parameters, is based on a finite element analysis (FEA) (Kobayashi et al. 1989). A variety of presses (e.g., mechanical or hydraulic) can be used, and nitrogen cylinders are typically placed under the die as a cushion to absorb the energy of the punch stroke and provide blank holding forces. For complex parts, a stamping line with several presses and dies is used. Dies can be quickly removed from the press to enable the production of a variety of different parts on the same press line. A more detailed discussion of the stamping process, and equipment used, is provided in Chap. 2.

Typical quality problems in sheet metal stamping include wrinkling (due to compressive stresses), tearing (due to tensile stresses) and springback (due to elasticity). These are illustrated in Fig. 1.5. If the binder force is too high, locally in a particular area of the die, then the flow of material into the die is restricted and tearing is likely to occur in that region. If the binder force, again locally in a particular region, is too low then excessive material flow can lead to wrinkling. Springback can be accounted for in the design of the die, as well as by varying the punch force during the stroke to set the part geometry.

Mechanical presses typically allow the operator to set the desired stroke, speed and blank holder force. The blank holder force typically cannot be adjusted around the die, and it is not possible to control the binder or punch force during the short time in which the part is formed. Hydraulic presses, depending on the design, do provide some additional flexibility in setting the blank holder and punch forces during operation. Thus, the control capabilities of stamping presses is limited, and they require significant trial and error during the die try out process to establish the best settings for use in a production run. Furthermore, during production, factors such as blank material thickness and formability variations, and changes in lubrication, can act as disturbances and lead to quality problems.

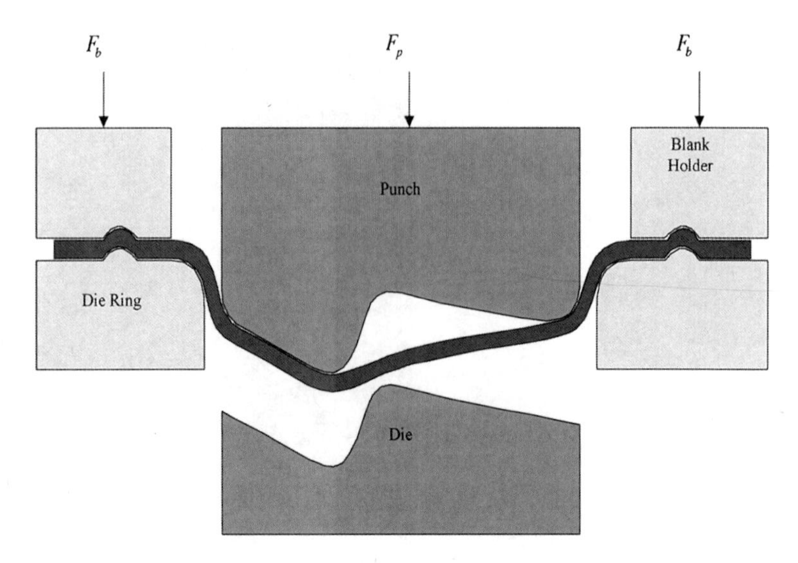

Fig. 1.4 Typical stamping press configuration, showing the punch, die and blank holder

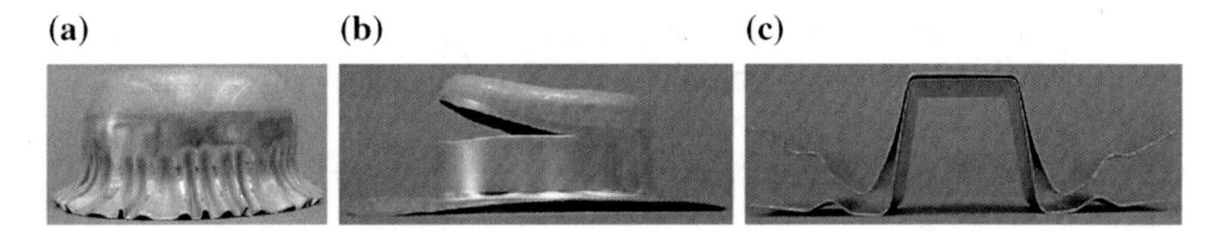

Fig. 1.5 Quality problems in stamping: **a** wrinkling, **b** tearing, and **c** springback

1.4 Stamping Process Control

A die try-out process, typically carried out over several days, is used to determine the final die geometry (e.g., by refining the die via grinding and welding) and to select the process conditions (e.g., tonnages and speed) that will produce good parts. This process requires experienced press operators and can be time consuming and costly. Furthermore, when the die is used in production it may be susceptible to process variations, such as variability in the sheet metal (e.g., formability or sheet thickness) or in the operating conditions (e.g., lubrication). Such process disturbances can lead to high scrap rates, even for a die that has been thoroughly vetted in the try out process. Consequently, there is a need for a process controller, as illustrated in Fig. 1.1, to provide consistent performance in the presence of disturbances. The state of the art in research on stamping process control is detailed in Chap. 3.

The key to effective process control in stamping is to be able to locally control the flow of material into the die (i.e., draw-in) during the press stroke (Hardt 1993).

This is a challenge with current presses, especially mechanical presses, since the binder forces cannot be locally controlled during the stroke. It is also not clear what process output (see Fig. 1.1) should be measured and fed back as an indicator of process performance. Ideally, the process output should be the material draw-in at various locations around the die, but that is a difficult quantity to measure. Furthermore, it is clear that a single feedback loop, as shown in Fig. 1.1, would not be sufficient for typical parts, due to their complex geometries. Instead one would need to measure the draw-in at various locations around the die, and also adjust the blankholder forces at multiple locations, to insure the right draw-in at those locations. Consequently, for effective process control of complex parts, a multi-input and multi-output (MIMO) process control system is required. It is also challenging to provide actuators that can locally control the material flow. Such a capability, which we term machine control, is essential as an inner-loop control to implement a process control loop as in Fig. 1.1. Chapter 4 provides a detailed discussion of the machine control problem for stamping.

In work done on a hydraulic laboratory press, a process control system was developed and validated based upon punch force measurement and subsequent adjustment of blank holder forces (Hsu et al. 2000, 2002). The process controller was based on a process model, similar to Eq. (1.1), with punch force, F_p, as the output and blank holder force, F_b, as the input (Hsu et al. 2000). During the short duration of the stroke (e.g., <1 s) the blank holder force, F_b, is adjusted as a function of time to achieve a desired reference punch force, R, through feedback of the measured punch force, F_p. The reference punch force, R, is a function of time (or stroke) and is predetermined off-line to ensure good part quality (i.e., elimination of wrinkling, tearing and springback). While this laboratory stamping process controller was found to be very effective, it could only be used for simple parts, where the material draw-in is uniform around the die, because it is a single-input single-output (SISO) feedback control system (Hsu et al. 2002). This laboratory SISO stamping process controller is described in more detail in Chap. 5.

More recently, a MIMO version of this stamping controller has been developed and validated in both die try out and production tests (Lim et al. 2010, 2012). Furthermore, this MIMO process controller for stamping is designed to be reconfigurable, through the use of multiple hydraulic actuators. As shown in Fig. 1.6, these hydraulic actuators can replace the nitrogen cylinders typically used to cushion the die in the press. Furthermore, the number and location of these cylinders can be varied to accommodate different die geometries while providing the ability to locally control blank holder forces at each cylinder location. As will be described in detail in Chaps. 6–8, this system also measures punch forces, at four corners of the press, and then adjusts blank holder forces at each hydraulic cylinder to follow predetermined reference values of the measured punch forces. The system has been experimentally evaluated in extensive die try out and production tests, with excellent results (Lim et al. 2010, 2012). Problems of wrinkling, tearing and springback are eliminated. It has been shown to be effective for a variety of sheet metal materials, including aluminum and lightweight alloys. The system is designed to be easy to use by press operators with minimal training.

Fig. 1.6 Reconfigurable design of hydraulic actuators supporting a die in a stamping press

Typical die try-out times are reduced from days to hours, and without the need for welding and grinding of dies. In the production environment, the system is capable of eliminating the effects of disturbances (e.g., variations in lubrication and/or material thickness), providing consistent part quality and reducing scrap rates.

The remainder of the book expands on the brief overview in this chapter, and provides a detailed engineering discussion and evaluation of this MIMO variable binder force process control system for stamping.

1.5 Purpose and Scope

The purpose of this book is to describe in detail a process control system for sheet metal stamping. In this chapter process control in manufacturing, and specifically in stamping processes, has been introduced. The important economic role of manufacturing, and of manufacturing automation in particular, has been highlighted and a historical perspective provided. The stamping process, and the potential benefits of process control, have been briefly described. In a stamping process control system (see Fig. 1.1) appropriate indicators of process performance (e.g., punch forces) are measured, and then used to adjust the process (e.g., blank holder forces) to achieve consistency (e.g., eliminate tearing, wrinkling and springback) in the presence of disturbances such as blank material variations.

The next chapter provides an expanded description of the stamping process, and Chap. 3 reviews recent research advances in control of sheet metal stamping (Lim et al. 2008). In Chap. 4, through the introduction of reconfigurable hydraulic actuators, a machine control system to achieve specified binder forces is presented. In Chap. 5 an additional process control level for stamping is introduced for a SISO laboratory system (Hsu et al. 2000, 2002). Chapter 6 presents detailed design and evaluation of a MIMO process controller for sheet metal stamping (Lim et al. 2010) and an adaptive version of this controller is then given in Chap. 7 (Lim et al. 2012). In Chap. 8 two adaptive control schemes (i.e., direct and indirect) are compared, and Chap. 9 provides some concluding remarks. References are provided in each chapter.

References

Cohen, S. S., Zysman, J. (1987) Manufacturing matters: the myth of the post-industrial economy. Basic Books, New York

Cohen, S. S., Zysman, J. (1988) Manufacturing innovation and American industrial competitiveness. Science 239(4844): 1110–1115

Hardt, D. E. (1993) Modeling and control of manufacturing processes: getting more involved. ASME J. of Dynamic Systems, Measurement and Control 115(2(B)): 291–300

Hosford, W. F., Caddell, R. (2011) Metal forming: mechanics and metallurgy. Cambridge University Press, Cambridge.

Hsu, C. W., Ulsoy, A. G., Demeri, M. Y. (2000) An approach for modeling sheet metal forming for process controller design. ASME J. of Manufacturing Science and Engineering 122(4): 717–724

Hsu, C. W., Ulsoy, A. G., Demeri, M. Y. (2002) Development of process control in sheet metal forming. J. of Materials Processing Technology 127(3): 361–368

Kalpakjian, S., Schmid, S.R. (2001) Manufacturing engineering and technology. Prentice-Hall, New Jersey

Kobayashi, S., Oh, S. I., Altan, T. (1989) Metal forming and the finite-element method. Oxford University Press, Oxford.

Koren, Y. (1983) Computer control of manufacturing systems. McGraw-Hill, New York

Koren, Y. (2010) The global manufacturing revolution: product-process-business integration and reconfigurable systems. Wiley, New York.

Koren, Y., Heisel, U., Jovane, F., Moriwaki, T., Pritchow, G., Ulsoy, A. G. and VanBrussel, H. (1999) Reconfigurable manufacturing systems. CIRP Annals 48(2): 527–540

Lauderbaugh, L.K., Ulsoy, A. G. (1988) Dynamic modeling for control of the milling process. ASME Journal of Engineering for Industry, 110(4): 367–375

Lauderbaugh, L. K., Ulsoy, A. G. (1989) Model reference adaptive force control in milling. ASME Journal of Engineering for Industry, 111(1): 13–21

Lim, Y., Venugopal, R., Ulsoy, A. G. (2008) Advances in the Control of Stamping Processes. Proc. IFAC World Congress, 17(1): 1875–1883

Lim, Y., Venugopal, R., Ulsoy, A. G. (2010) Multi-input multi-output (MIMO) modeling and control for stamping. ASME J. Dynamic Systems, Measurement and Control 132(4): 041004 (12 pages)

Lim, Y., Venugopal, R., Ulsoy, A. G. (2012) Auto-tuning and adaptive stamping process control. Control Engineering Practice 20(2): 156–164

Marciniak, Z., Duncan, J. L., Hu, S. J. (2002) Mechanics of sheet metal forming. Butterworth-Heinemann, Oxford.

Rashap, B. A., Elta, M. E., Etemad, H., Fournier, J. P., Freudenberg, J. S., Giles, M. D., Grizzle, J. W., Kabamba, P. T., Khargonekar, P. P., Lafortune, S., Moyne, J. R., Teneketzis, D., Terry, F. L. (1995) Control of semiconductor manufacturing equipment: real-time feedback control of a reactive ion etcher. IEEE Trans. on Semiconductor Manufacturing 8(3): 286–297

Ulsoy, A. G., Koren, Y., Rasmussen, F. (1983) Principal developments in the adaptive control of machine tools. ASME Journal of Dynamic Systems, Measurement and Control 105(2): 107–112

Ulsoy, A. G., Koren, Y. (1993) Control of machining processes. ASME Journal of Dynamic Systems, Measurement and Control 115(2(B)): 301–308

Womack, J. P., Jones, D. T., Roos, D. (1990) The machine that changed the world. HarperPerrenial, New York

Chapter 2
Equipment and Material Flow Control

Abstract Process control ensures that stamped part quality is maintained in the presence of operational variations and disturbances. Control objectives are achieved by adjusting the flow of metal into the die cavity in response to these variations and disturbances. This chapter provides an overview of the process and equipment used for sheet-metal stamping, and describes the methods through which material flow control can be effected during the stamping process. Simplified kinematic and dynamic models for press motion are derived, followed by a description of hydraulic actuation to implement variable binder force for material flow control.

2.1 Types of Stamping Presses

The sheet metal stamping process, described in Sect. 1.3, will now be explored in greater detail. We begin with an overview of the two main types of stamping presses that are commonly used in the manufacture of deep-drawn parts, namely, *mechanical presses* and *(electro-) hydraulic presses*.

Metal-forming presses are used for a number of operations including deep-drawing, blanking and trimming. A typical *press-line* in a production facility has several presses in a line, each of which performs one or more operations on a part, with automation sequentially moving the part along the line. Our study will be confined to presses used for drawing.

As seen in Fig. 1.4, for drawing, the punch has to be driven by a mechanism to force the blank into the die. Mechanical presses use a linked-drive powered by an electric motor to drive the punch. A clutch mechanism is used to engage or disengage the drive, and a braking system is included to stop the drive. Hydraulic presses use hydraulic cylinders to drive the punch. Servo-valves or proportional valves are utilized to control the flow of pressurized hydraulic fluid into the cylinders during the punch stroke.

Y. Lim et al., *Process Control for Sheet-Metal Stamping*,
Advances in Industrial Control, DOI: 10.1007/978-1-4471-6284-1_2,

The two main characteristics of a press that affect the forming process are the *press tonnage*, which is the maximum force applied by the press on the blank (typically at the bottom of the forming stroke) and the *press curve*, which is the variation of the speed of the punch during the forming stroke. Most modern presses have sensors, or load-cells, called *tonnage monitors*, which measure the press tonnage at the four corners of the press. In a mechanical press, the press curve is determined by the kinematics of the drive mechanism, while for a hydraulic press, the punch speed is constant until the punch slows down close to the bottom of the stroke. However, modern *electro-hydraulic servo-presses* allow the operators to program desired press curves to enable better forming control.

The rate at which a press runs, specified in *strokes per minute*, determines the production rate, while the *shut-height* of the press, which is the vertical displacement of the punch as it impacts the blank, determines the size of the die that can be accommodated in a press for a given draw depth.

Presses, both mechanical and electro-hydraulic can be either *single-acting* or *double-acting*. In a single-acting press, the drive mechanism only moves the press ram, while in a double-acting press, the drive mechanism moves an outer blank-holding ring which clamps the blank before it drives the press ram to form the metal. Variable binder-force control is more effective in single-acting presses, and thus, our discussion in this book will be limited to these presses.

2.2 Mechanical Presses

The basic functional components in a mechanical press are shown in Fig. 2.1. The drive mechanism consists of an electric motor connected to a crank mechanism to move the punch along a guide on the sides of the press-frame (Adam et al. 1998). We now develop a simplified model of the kinematics of a mechanical press.

Referring to Fig. 2.2, we see that

$$y(t) = R\cos(\theta(t)) + \sqrt{l^2 - R^2\sin^2\theta(t)} \tag{2.1}$$

and differentiating (2.1) with respect to time yields

$$\dot{y}(t) = -R\sin(\theta(t))\omega - \frac{R^2\sin(2\theta(t))\omega}{2\sqrt{l^2 - R^2 sin^2\theta(t)}} \tag{2.2}$$

where t denotes time and $\omega = \dot{\theta}(t)$ denotes the constant angular speed of the press-drive, with a super-scripted dot indicating the time-derivative of a time-dependent function.

Figure 2.3 shows the simulated press-curve of the press-drive modeled above, with $\omega = 2.0944$ rad/s, corresponding to a press running at 20 strokes/min, and $R = 0.5$ m, $l = 2.5$ m. Note that the speed of punch decreases rapidly as the punch reaches the bottom of the stroke, ensuring that the final stages of forming are achieved at a lower strain rate.

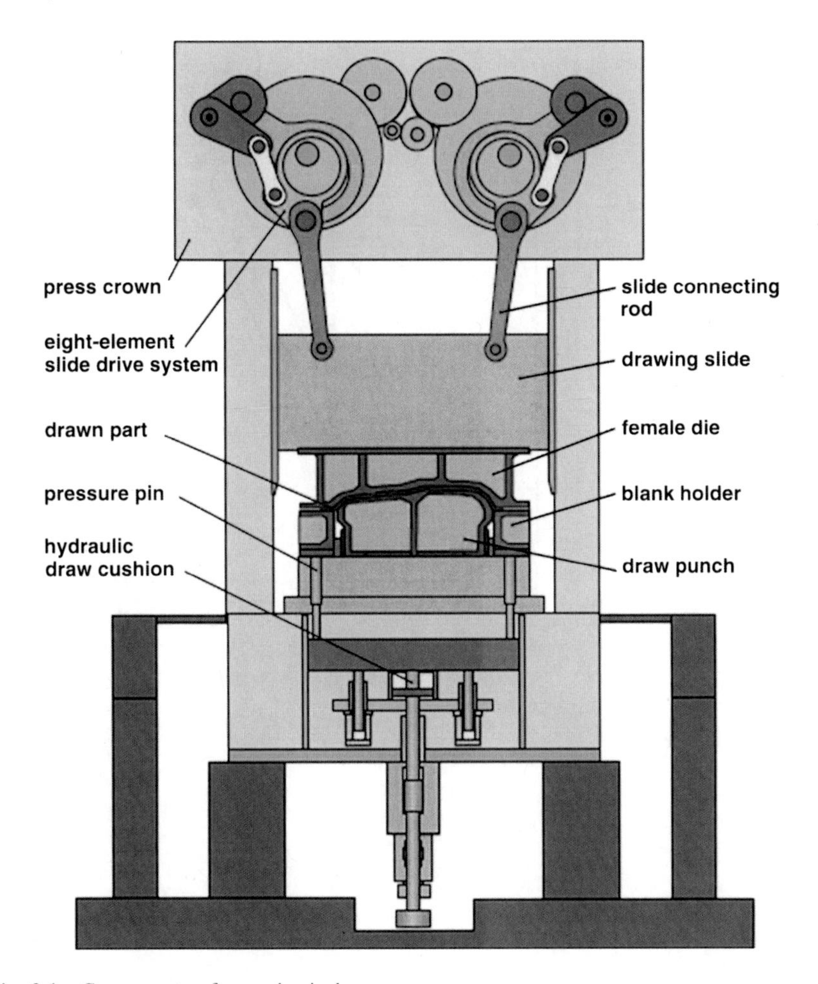

Fig. 2.1 Components of a mechanical press

2.3 Hydraulic Presses

Hydraulic (or electro-hydraulic) presses are also used for deep-drawing sheet metal. In contrast to mechanical presses, the driving force of the press is generated by pressurized hydraulic fluid. The press ram is driven by hydraulic cylinders attached to the press ram as shown in Fig. 2.4.

A simplified mathematical model to derive the press ram-speed, $\dot{y}(t)$, as a function of the supply pressure, P_s, the bulk-modulus of the hydraulic fluid, β, the density of the hydraulic fluid, ρ, the cross-sectional areas of the hydraulic cylinder in contact with the fluid on the rod-side and the blind-side, A_1 and A_2 respectively, the servo-valve effective flow-area, K_v, and the total length of the cylinder, l, is shown below.

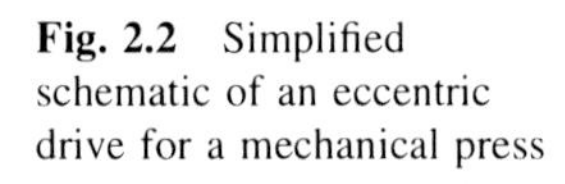

Fig. 2.2 Simplified schematic of an eccentric drive for a mechanical press

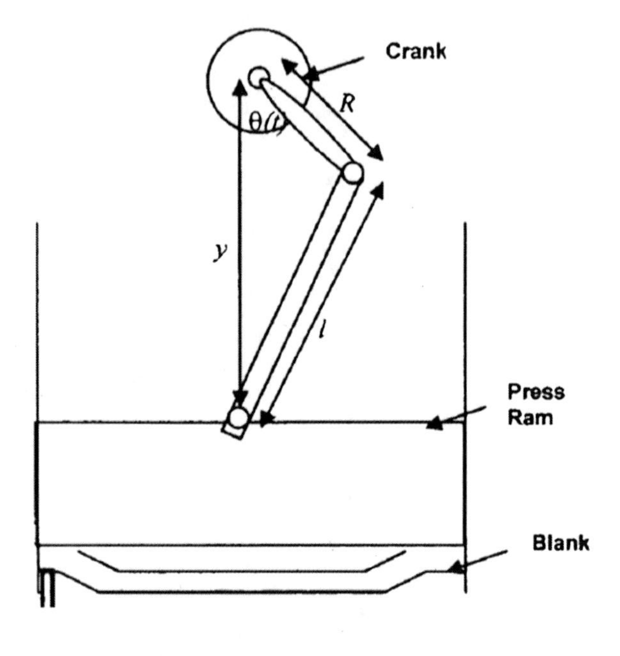

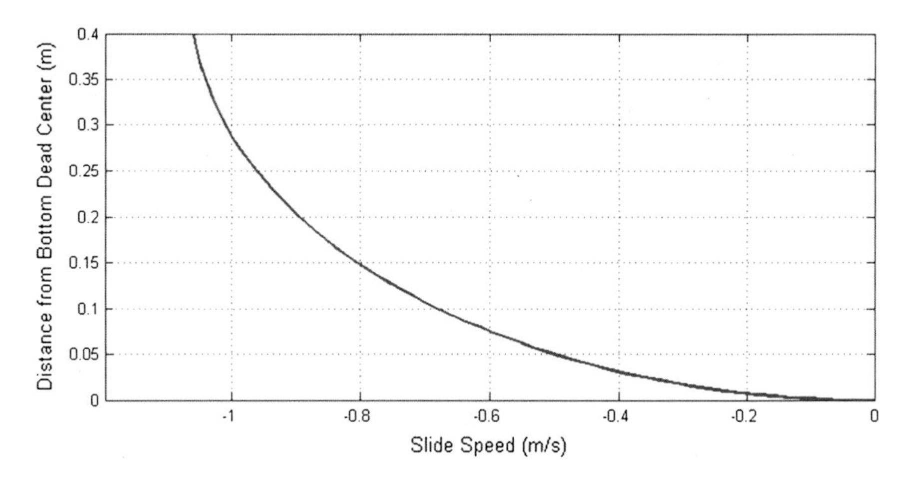

Fig. 2.3 Simulated mechanical press curve

Referring to the simplified schematic in Fig. 2.5, the pressurized hydraulic fluid enters one side of the double-acting hydraulic cylinders via servo-valves. The return flow from the other side of the cylinders is handled by the same servo-valves.

First, the change in pressure in a hydraulic fluid, ΔP, is related to the bulk-modulus, volumetric change, ΔV, and the instantaneous volume of the hydraulic fluid, $V(t)$, by (Merrit 1991)

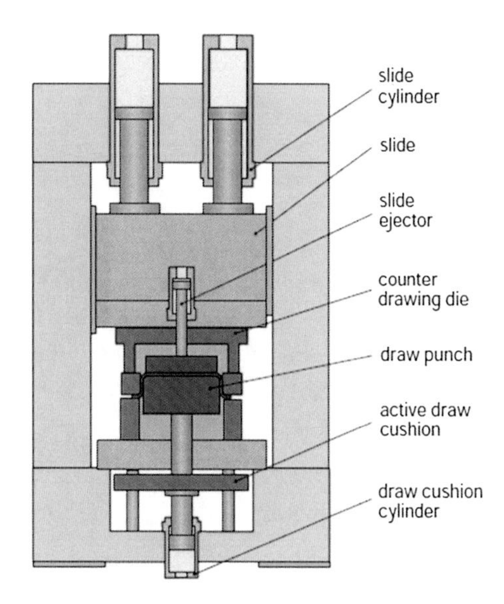

Fig. 2.4 Components of a hydraulic press

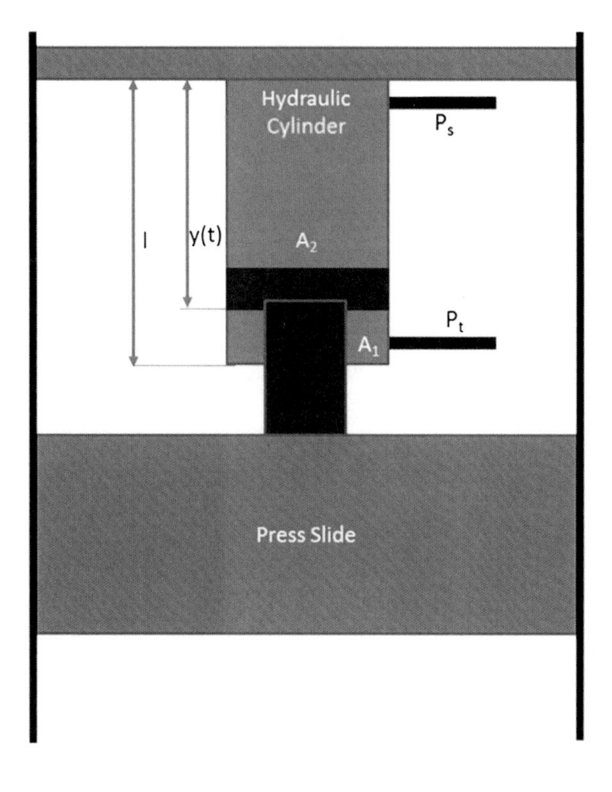

Fig. 2.5 Simplified schematic of hydraulic press drive

$$\Delta P = -\beta \frac{\Delta V}{V(t)} \tag{2.3}$$

or,

$$\dot{P}(t) = -\beta \frac{\dot{V}(t)}{V(t)}. \tag{2.4}$$

Denoting the pressure of the hydraulic cylinder on the rod-side as P_1 and that on the blind-side as P_2, assuming laminar flow and using Bernoulli's equation, the rate of hydraulic fluid flow into the blind-side of the cylinder, $Q_2(t)$, is given by

$$Q_2(t) = K_v \sqrt{\frac{2(P_s - P_2)}{\rho}} \tag{2.5}$$

and thus, the net rate of change of volume of the hydraulic fluid in the blind-side, $\dot{V}_2(t)$, taking into account the vertical motion of the piston, is

$$\dot{V}_2(t) = Q_2(t) - A_2 \dot{y}(t) \tag{2.6}$$

or,

$$\dot{V}_2(t) = K_v \sqrt{\frac{2(P_s - P_2)}{\rho}} - A_2 \dot{y}(t). \tag{2.7}$$

Similarly, using analogous notation for the rod-side, it can be shown that

$$\dot{V}_1(t) = A_1 \dot{y}(t) - K_v \sqrt{\frac{2(P_1)}{\rho}} \tag{2.8}$$

assuming that the tank pressure is 0.

Next, using (2.4), (2.7) and (2.8), it follows that

$$\dot{P}_1(t) = \beta \frac{A_1 \dot{y}(t) - K_v \sqrt{\frac{2(P_1)}{\rho}}}{A_1(l - y(t))} \tag{2.9}$$

and

$$\dot{P}_2(t) = \beta \frac{K_v \sqrt{\frac{2(P_s - P_2)}{\rho}} - A_2 \dot{y}(t)}{A_2 y(t)}. \tag{2.10}$$

Finally, letting m denote the mass of the slide and punch and invoking Newton's Second Law of Motion, it follows that

$$m\ddot{y}(t) = P_2 A_2 - P_1 A_1. \tag{2.11}$$

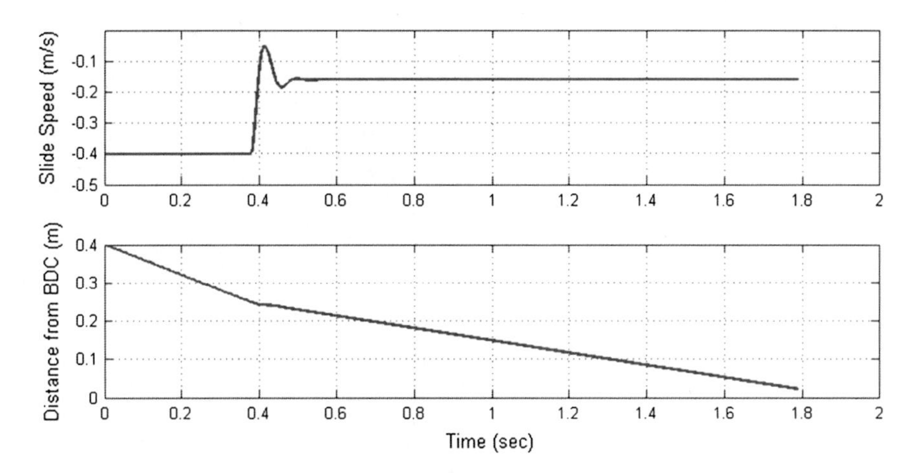

Fig. 2.6 Simulated slide speed profile for a hydraulic press

The press ram-speed, $\dot{y}(t)$ can thus be calculated by numerically integrating Eqs. (2.9), (2.10) and (2.11).

Figure 2.6 shows a simulated slide speed profile for a hydraulic press. The press parameters are as follows: $m = 50$ tons, $P_s = 35$ MPa, $A_1 = 0.15\ \text{m}^2$, $A_2 = 0.5\ \text{m}^2$, $\beta = 1.25$ GPa, $l = 2.5$ m and $\rho = 850\,\text{kg/m}^3$. The servo-valve is rated at 1,500 l/min at a pressure differential of 0.5 MPa. In contrast to a mechanical press, the speed of the slide can be controlled by regulating the flow of hydraulic fluid into the cylinder using the servo-valve. For a given setting of the spool of the servo-valve, the speed of the cylinder reaches a constant speed. In the upper plot, it can be seen that the slide is lowered at a speed of 0.4 m/s until the ram comes close to contact with the blank and then it is slowed down to a speed of 0.18 m/s for the forming cycle. The corresponding slide displacement is shown in the lower plot. The models derived above to characterize the slide speeds for both mechanical and hydraulic presses are useful in the design of hydraulically actuated variable binder force systems. The binder force is generated by compression of the hydraulic fluid as the binder moves down, and the speed of the binder is the same as the slide speed once the ram comes in contact with the die.

2.4 Variable Binder Force Control

As described in Sect. 1.3, binder force plays an important role in how the material flows during the forming process, and is thus crucial to part quality. Two methods are conventionally used to control material flow; the first utilizes drawbeads which are designed using FEA, while the second uses fixed binder forces applied either using a *draw cushion* or nitrogen cylinders (*nitros*). With both these methods, the material flow cannot be actively changed during the forming process.

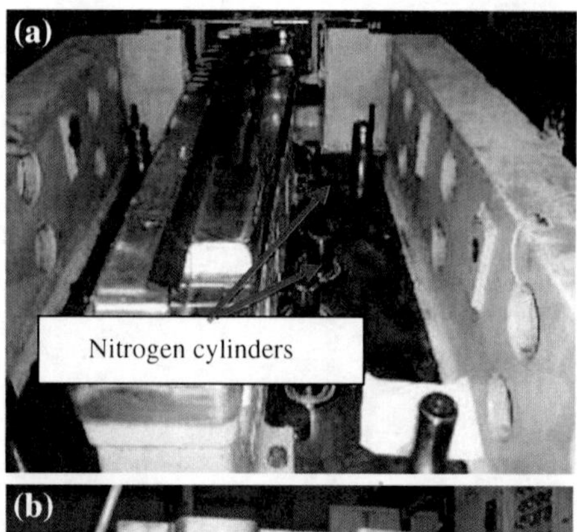

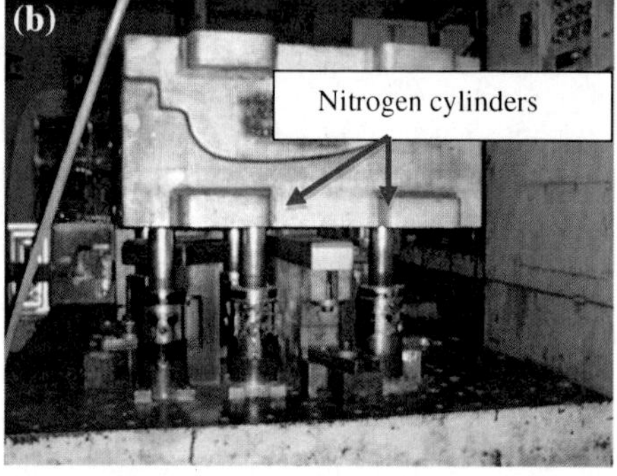

Fig. 2.7 **a** Production-out die with nitrogen cylinders installed, **b** try-out die with nitrogen cylinders installed

A draw cushion generates binder force using compressed air or hydraulic fluid located in the bed of the press underneath the die (see Figs. 2.1 and 2.4). The force is transmitted by a number of cushion pins to the binder. As the pressure in the fluid is constant across all pins, the binder force cannot be varied spatially. Furthermore, the pressure in the fluid builds up during the stroke and is typically not controlled. Newer hydraulic press cushions are sometimes segmented into quadrants and allow pressure variation in preset steps, allowing a limited amount of binder force control in both space and time, but these adjustments are typically coarse.

Nitrogen cylinders are passive cylinders with pressurized nitrogen, which are mounted under the binder. Figure 2.7 shows production and try-out dies with nitrogen cylinders installed. The pressure in each cylinder can be adjusted locally to vary binder force, but the force in these cylinders always increases during the stroke, due to compression of the gas. Thus, the total binder force at the end of the stroke, where splits typically occur, is always higher than at the start of the stroke, resulting in limited ability to affect material flow during the stamping cycle.

Extensive research has been conducted on the use of variable binder force for part quality improvement (Hashida and Wagoner 1993; Siegert et al. 1997, 2000; Krishnan and Jian 2003; Hsu et al. 2000, 2002; Lim et al. 2008, 2010, 2012). The most effective method is to vary the binder force locally at a number of points around the die (typically 12–20) (Lim et al. 2010, 2012) as lightweight materials formed in complex shapes have a tendency to have both splits and wrinkles in different areas.

An effective way of generating binder force that can be varied both spatially and temporally during the stroke is to use a set of hydraulic cylinders. The pressure within each individual cylinder is controlled during the press stroke using high-bandwidth servo-valves. For example, using such a system, the binder force in one location can be maintained at a relatively high level for most of the stroke and then relaxed in the last 1 cm of stroke to prevent a split. At another location, the binder force can be increased in the last 1.5 cm of stroke to reduce spring-back or to create the desired stretch. In Chap. 4, the design of hydraulic actuation systems for variable binder force control will be treated in detail.

It is noted that for local variation of binder force to be effective, the binder has to flex. Several studies have been conducted on the use of flexible binders for improving part quality (Siegert et al. 1997, 2000; Krishnan and Jian 2003). Increased flexibility is achieved through the use of segmented binders. However, more recent studies have shown that standard "rigid" binders show sufficient flexibility to ensure local material flow control with locally varying binder force, in both die try-out and production (Lim et al. 2010, 2012).

Alternative methods of actively controlling material flow during the stamping process include the use of active drawbeads (Li and Weinmann 1999; Bohn et al. 2001) ultrasonic vibrations (Takashi et al. 1998) and electromagnetic impulses (Daehn et al. 1999; Shang and Daehn 2011). With active drawbeads, the bead is segmented and the height of each segment can be adjusted during the stamping cycle to control material flow. Ultrasonic vibrations can also be locally targeted to vary the friction between the blank and the die surface during the forming process, thus allowing local material flow control. Finally, electromagnetic impulses directly affect the strain energy in the material. While all these methods have shown the ability to improve part quality in laboratory studies, they all involve complex actuation mechanisms and their implementation on the shop is currently prohibitively expensive.

2.5 Overview of Machine and Process Control

To close this chapter, the overall control architecture for stamping process control is reviewed. In Sect. 2.4, actuation methods to effect material flow control were described. Each of these methods requires a closed-loop to ensure that the actuator changes the control variable, for example, local binder-force or active draw-bead height, as desired. Once this can be done, an outer process-control loop is used to generate reference commands for the control variable to achieve the desired process-control objectives. The block-diagram in Fig. 2.8 shows the control architecture using hydraulic actuators for variable binder force.

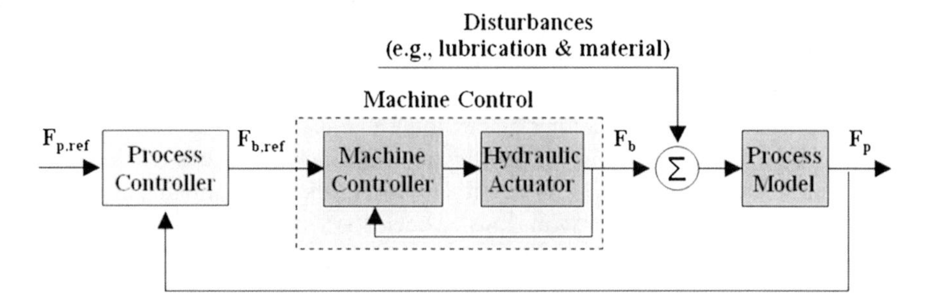

UNUSED C

Cyl Stroke (in)	Cyl 1 Tonnage	Cyl 2 Tonnage	Cyl 3 Tonnage	Cyl 4 Tonnage	Cyl 5 Tonnage	Cyl 6 Tonna
0.00	15.0	15.0	15.0	15.0	15.0	15.0
0.25	15.0	15.0	15.0	15.0	15.0	15.0
0.50	15.0	15.0	15.0	15.0	15.0	15.0
0.75	15.0	15.0	15.0	15.0	15.0	15.0
1.00	15.0	15.0	15.0	15.0	15.0	15.0
1.25	15.0	15.0	15.0	15.0	15.0	15.0
1.50	15.0	15.0	15.0	15.0	15.0	15.0
1.75	15.0	15.0	15.0	15.0	15.0	15.0
2.00	15.0	15.0	15.0	15.0	15.0	15.0
2.25	15.0	15.0	15.0	15.0	15.0	15.0
2.50	15.0	15.0	15.0	15.0	15.0	15.0

Fig. 2.8 Control architecture for stamping process control

Fig. 2.9 Binder tonnage input user interface

The inner actuation control is achieved using a *machine controller*. All further discussion will be limited to variable binder force systems utilizing hydraulic actuators. Formability can be improved by the use of machine control alone (Lim et al. 2010); a skilled operator manually inputs binder force reference commands ($F_{b,ref}$) for each actuator to correct splits, wrinkles or spring-back. Figure 2.9 shows an input screen for machine control of a commercial 12-cylinder variable binder force system. The binder force applied by each actuator is commanded in 6.35 mm (0.25 inch) increments of punch travel during the forming cycle. The machine controller ensures that these binder force commands are tracked by the actuator and delivered to the binder. The best binder force profiles to make a part of desired quality under nominal conditions are established during try-out. Initial estimates for variable binder force profiles may be generated using FEA prior to try-out (Ahmetoglu et al. 1995; Sheng et al. 2004).

These force profiles create a baseline under nominal conditions. However, operational variations such as changes in lubrication or material properties can result in part defects, if the binder force profiles are maintained at their nominal settings. When such defects occur, the die try-out specialists may be able to adjust the binder force profiles to compensate for the operational variables but such changes are typically made using trial-and-error and may result in unacceptable downtime during a production run. A more robust approach is to use a process control to continually monitor a process variable that reflects part quality and automatically correct the binder force reference commands fed to the machine controller. These corrections modify the nominal binder force commands to ensure that part quality is maintained in the face of operational variations.

The hydraulic dynamics of the actuators induce nonlinearities that need to be addressed in the machine controller. Chapter 4 describes the use of feedback linearization to handle these nonlinearities. The fact that multiple actuators are used with multiple process variable sensors leads to the need to design a multiple-input multiple-output (MIMO) process controller. Furthermore, the number of sensors is usually much lower than the number of actuators, leading to a non-square MIMO structure. The process dynamics are also nonlinear; however, experimental tests indicate that linearized models are sufficient to correct typical part defects that occur in production runs (Lim et al. 2010, 2012). Chapters 6, 7 and 8 describe several methods for process control design, starting with a simple PID approach and moving to direct and indirect adaptive control to handle the large uncertainty in parameters.

References

Adam, K. et al. (1998) Schuler metal forming handbook. Springer-Verlag, Berlin.

Ahmetoglu, M., Broek, T. R., Kinzel, G., Altan, T. (1995) Control of blank holder force to eliminate wrinkling and fracture in deep-drawing rectangular parts. CIRP Annals 44(1): 247–250

Bohn, M., Xu, S., Weinmann, K., Chen, C., Chandra, A. (2001) Improving formability in sheet metal stamping with active drawbead technology. J. of Engineering Materials and Technology 123: 504–510.

Daehn, G. S., Vohnout, V. J., DuBois, L. (1999). Improved formability with electromagnetic forming: fundamentals and a practical example. Minerals, Metals and Materials Societty/ AIME, Sheet Metal Forming Technology (USA): 105–115.

Hashida, Y., Wagoner, R. H. (1993) Experimental analysis of blank holding force control in sheet forming, SAE Paper 930285, Sheet Metal Stamping Symposium, SAE SP-994: 93–100, Warrendale, PA.

Hsu, C. W., Ulsoy, A. G., Demeri, M. Y. (2000) An approach for modeling sheet metal forming for process controller design. ASME J. of Manufacturing Science and Engineering 122(4): 717–724

Hsu, C. W., Ulsoy, A. G., Demeri, M. Y. (2002) Development of process control in sheet metal forming. J. of Materials Processing Technology 127(3). 361–368

Krishnan, N., Jian, C. (2003). Estimation of optimal blank holder force trajectories in segmented binders using an ARMA model. Journal of manufacturing science and engineering, 125(4): 763–770

Li, R.i, Weinmann, K. J. (1999) Non-symmetric panel forming of AA 6111-T4 using active drawbeads, Proc. of TMS symposium, San-Diego, p.39–52

Lim, Y., Venugopal, R., Ulsoy, A. G. (2008) Advances in the Control of Stamping Processes. Proc. IFAC World Congress, 17(1): 1875–1883

Lim, Y., Venugopal, R., Ulsoy, A. G. (2010) Multi-input multi-output (MIMO) modeling and control for stamping. ASME J. Dynamic Systems, Measurement and Control 132(4): 041004 (12 pages)

Lim, Y., Venugopal, R., Ulsoy, A. G. (2012) Auto-tuning and adaptive stamping process control. Control Engineering Practice 20(2): 156–164

Merritt, H. E. (1991) Hydraulic control systems. Wiley, New Jersey

Shang, J., and Daehn, G. (2011). Electromagnetically assisted sheet metal stamping. Journal of Materials Processing Technology, 211(5): 868–874

Sheng, Z. Q., Jirathearanat, S., Altan, T. (2004) Adaptive FEM simulation for prediction of variable blank holder force in conical cup drawing. Int. J. of Machine Tools and Manufacture 44(5): 487–494

Siegert, K., Altan, T., Nakagawa, T. (1997). Development and manufacture of dies for car body production. CIRP Annals-Manufacturing Technology, 46(2): 535–543.

Siegert, K., Häussermann, M., Haller, D., Wagner, S., Ziegler, M. (2000). Tendencies in presses and dies for sheet metal forming processes. Journal of Materials Processing Technology, 98(2): 259–264

Takashi, J., Yukio, K., Nobuyoshi, I., Osamu, M, Eiji M., Katsuhiko, I., Hajime, H. (1998) An application of ultrasonic vibration to the deep drawing process, Journal of Materials Processing Technology 80–81: 406–412

Chapter 3
Recent Advances in Stamping Control

Abstract This chapter presents a review of research on control of the sheet metal stamping process, and its effect on the quality of stamped parts. First the evolution of control strategies for the forming process is presented. Next the different types of active blank holder force systems from previous research are described. Finally, a review of in-process sensor technologies to monitor the process variables used in process controllers for sheet metal stamping is given.

3.1 Developments in Stamping Control

Current research has focused on four possible ways to tackle the challenges in sheet metal forming. First, die try-out with open-loop control is a pre-process procedure to adjust tooling designs and process variables. Next, closed-loop machine control is the on-line control of blank holder force (BHF) based on finite element analysis (FEA) simulation or die-maker experience. Thirdly, in-process control is an on-line strategy to apply feedback control to process inputs (e.g., BHF and drawbead restraining force) to mitigate the effect of disturbances during the forming process. Finally, cycle-to-cycle control is based on post-process part inspection to determine critical process variables to be monitored.

3.1.1 Die Try-Out

Die try-out determines the parameters (such as die geometry and BHF) that control the forming process to avoid tearing and wrinkling by physical modifications (e.g., grinding and welding) of the die surface and alteration of the BHF in sheet metal stamping. The die try-out procedure is time-consuming, with many cycles of trial and error. Sklad and Harris (1991) noted that most changes in stamping are

Y. Lim et al., *Process Control for Sheet-Metal Stamping*,
Advances in Industrial Control, DOI: 10.1007/978-1-4471-6284-1_3,

connected with altering die geometry data (e.g., binder geometry and drawbeads), because part geometry data and material data are normally fixed in the stamping process. They also noted that a poorly-tuned open-loop forming process, which is close to failure, can result in frequent disruptions in manufacturing scheduling and a high scrap rate, and significantly increase costs. Therefore, Finite Element Method (FEM) software tools play an important role in rapid evaluation of forming severity, with respect to fracture and wrinkling, prior to actual die manufacturing in order to reduce costs and scrap rate.

Although die try-out is costly and time-consuming, many techniques have been incorporated in die try-out, and lead-time and production costs have been improved. Herderich (1990) has developed empirical equations to predict the force necessary to form sheet metal around drawbeads. He suggested useful concepts in determining BHF and the number of nitrogen gas cylinders and/or nitrogen gas pressures with respect to quality of stamped parts. Xu and Zhao (2007) discussed the reduction of springback through open-loop compensation of mechanics-based springback reduction (e.g., drawbead constraint force) and geometry-based springback reduction (e.g., die face compensation).

3.1.2 Closed-Loop Machine Control

The closed-loop machine control strategy, as illustrated in Fig. 3.1, is to control the BHF, F_b, to follow a predetermined reference trajectory. Closed-loop machine control requires a predetermined reference BHF trajectory that can be obtained through experiment and/or FEM simulation. The reference BHF depends on its location on the die and is a function of the punch stroke. The punch force (F_P), as shown in Fig. 3.1, is an output of the process and is directly associated with splitting and wrinkling. Punch force is affected by the BHF. Thus, the relationship between the punch force and the BHF, obtained using mathematical modeling and/or experiments, can be used to design a process controller which generates a reference BHF trajectory for the machine controller. A detailed discussion of machine control is given in Chap. 4. In this section we briefly review research on machine control in stamping.

Different kinds of BHF trajectories (e.g., a step change in the BHF) have been used to experimentally study their effects on produced part quality. Some of them have been demonstrated to improve formability (Ahmetoglu et al. 1992; Kergen and Jodogne 1992; Ziegler 1999), to reduce springback (Adamson et al. 1996; Sunseri and Cao 1996; Siegert 1992, 1997; Ziegler 1999), and to improve part consistency (Adamson et al. 1996). However, these investigations have not led to methods to determine BHF trajectories required to make good parts.

Kergen and Jodogne (1992), however, showed how to experimentally design a minimum BHF trajectory and a maximum one using closed-loop machine control based on wrinkle detection. However, the measured BHF trajectories, and the

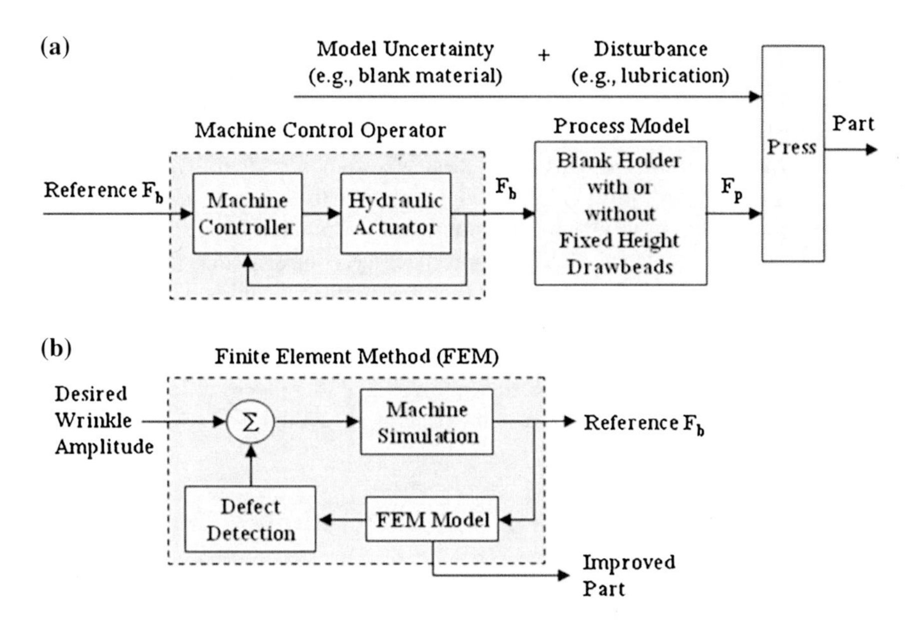

Fig. 3.1 Closed-loop machine control: **a** adjust the BHF to achieve the BHF, **b** determination of the BHF profile using FEM analysis

minimum BHF obtained from experiments varied significantly with the properties of the steels used in the experiments. The initial BHF values used by the authors were high, from which point the closed-loop control began to decrease the BHF, as the punch load increased, in order to ensure that the material is stretched to the highest level without causing tearing. The minimum BHF value corresponded to approximately the same point in the punch stroke as the maximum punch force. The authors found such a scheme yields improvements in the achievable limiting draw ratios.

Sunseri et al. (1996) showed how to determine the variable BHF reference trajectories by both finite element simulation and experiments, for winkling and tearing control. They monitored punch force history during the process, if the punch force history deviated from the target punch force trajectory, a proportional-integral (PI) controller acted to change the BHF during the stamping process in order to maintain the desired target punch force history based on the amounts of draw-in and thickness distribution of sheet metal. A FEM of the forming process was used to simulate the same controlled process. The corresponding BHF histories were obtained for different friction cases, in order to maintain the desired punch force.

Sheng et al. (2004) and Zhong-qin et al. (2007) predicted the optimal magnitude of the BHF to improve fracture and wrinkling problems in deep drawing. They used FEM simulation of closed-loop control based on the wrinkling and fracture detection of sheet metal. They showed the variable BHF profile predicted by adaptive FEM simulation, and compared the optimum constant BHF profile (see Fig. 3.1b). Using a pre-determined variable BHF profile, they formed a cup to a depth of 47 mm without any failures. Compared with a cup formed by optimum constant BHF, this represented an increase of 9 % in cup depth. Optimal trajectories of variable BHF by FEM simulation were developed in (Zhong-qin et al. 2007), and shown with a PID closed-loop controller to increase the forming limit by 30 % (i.e., from 45 to 60 mm forming depth) compared to constant BHF.

The main disadvantage of closed-loop machine control is that it cannot eliminate the influence of disturbances (e.g., variations in lubrication and blank thickness) on part quality and consistency. For example, two tests were conducted using closed-loop machine control with a predetermined BHF trajectory but with different lubrication conditions. Results showed that there were differences in part quality but not in BHF for the two tests (Hsu et al. 2000).

3.1.3 In-Process Control

As introduced previously (see Sects. 1.2, 1.4 and 2.5) in-process control is used to control a measurable process variable (e.g., punch force or draw-in) to follow a reference trajectory by manipulating the BHF (see Fig. 3.2). To implement the in-process control, a process controller and reference trajectory are needed after the monitored process variable is selected, in addition to the machine control scheme shown in Fig. 3.1. In this section we provide a brief review of research on process control in stamping.

Hardt and Fenn (1993) performed a series of constant BHF experiments to find failure height and then defined optimal tangential force (i.e., punch force) and normalized average thickness trajectories as the actual trajectories of these variables when the failure height was the largest. Then, they presented a method for in-process control of the BHF to ensure optimal forming conditions based on desired optimal trajectories. The method was implemented using closed-loop control based on process variable feedback, and subjected to experiments where various disturbances (e.g., lubrication and material change) were considered.

Siegert et al. (1997) showed that the material flow, or draw-in, is highly dependent on the friction force between the sheet metal and the upper and lower binder. They introduced process control using friction force as the controlled variable to avoid wrinkling and tearing during the stamping process. They also showed that the actual friction force follows the desired nominal curve of BHF. Therefore, they focused on monitoring the friction force by using a sensor, and utilized feedback control to realize the desired friction force curve over the stroke.

Bohn et al. (1998) developed a new multiple-point active drawbead forming die to improve part quality, using drawbead restraining force based on measuring the die shoulder force during the drawing process. In comparison to their previous

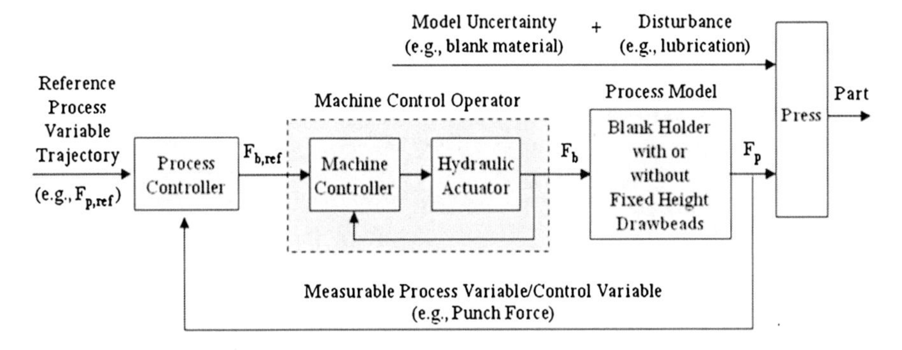

Fig. 3.2 In-process control of sheet metal forming process with reference punch force trajectory

work (Michler et al. 1994), they expanded the study to include multiple-point actuation with closed-loop control and developed second-order transfer functions for modeling the drawbead hydraulic actuators. They also monitored punch stretching force and adjusted the displacement of the active drawbead to constrain material flow, thus, avoiding tearing and wrinkling during the forming process.

Hsu et al. (2000, 2002) demonstrated that in-process control can be used to improve stamped part quality and consistency of a simple part by adjusting the BHF in forming based on tracking an optimal punch force trajectory. They pointed out that a process controller and reference punch force trajectory had to be included in the design (see Fig. 3.2). In particular, their approach included modeling of the sheet metal forming process, design of the process controller, and determination of the optimal punch force trajectory. They achieved good results using a proportional-integral controller with feedforward action (PIF) process controller in both simulation and experiments (see Chap. 5).

Kinsey et al. (2000) proposed a neural-network system, along with a stepped BHF trajectory, that was able to control springback in forming. A neural network was chosen due to its ability to handle the highly non-linear coupled effects that are found in sheet metal forming when variations in material and process parameters occur. Polynomial coefficients from curve fitting of the punch force trajectory were used as inputs into the neural network. Viswanathan et al. (2003) experimentally implemented the neural-network based process control for springback reduction during forming. They noted that neural-network control would be effective in dealing with material variations. However, for forming a complex part, they noted that more advanced sensors (e.g., local draw-in or local tangential force measurement) are needed because punch force alone is not sufficient in identifying variations.

Doege et al. (2003) described a new optical draw-in sensor for in-process material flow measurement and its application for closed-loop process control in sheet metal forming. They developed a press with a multi-point draw-in measurement tool within the control loop. They produced locally varying forces on the blank holder, in accordance with material flow information. Draw-in velocity along the drawing depth was controlled in accordance with desired BHF to produce parts without wrinkling and tearing.

Machine control can suffer from inconsistency when changes in lubrication and material properties occur. Thus, in-process control appears to be a reasonable solution for overcoming such production challenges in sheet metal forming. Although some success in applying process control to sheet metal forming has been reported, there are still many open questions. For example, the systematic design of the process controller and reference trajectory in forming processes have only recently been addressed and sensing and actuation technologies are not fully developed.

3.1.4 Cycle-to-Cycle Control

Statistical process control methods are used to implement cycle-to-cycle control based on the dimensional measurements of stamped parts. In cycle-to-cycle control, an important aspect is to maintain a database of process variables (e.g., material property, lubrication, BHF, punch force, draw-in, and punch speed). For example, as illustrated in Fig. 3.3, operator experience is necessary to adjust a process variable(s) at each cycle. Ultimately, the current cycle-to-cycle control, where an operator closes the loop using dimensional measurement in an otherwise open-loop process, could be improved when combined with machine control and in-process scheme.

Manabe et al. (1999) proposed the use of a database for an intelligent sheet metal forming system to enable design of a process control system without experts who are skilled and experienced in the forming process. They developed a fuzzy rule model, which provides an easy way to optimize cycle-to-cycle control, because the deep drawing process is not only unsteady and complicated but has nonlinear characteristics. Their method resulted in around 25 % reduction in production time. They were able to increase the draw-depth of an experimental cup by 0.77 mm using their method.

Hardt and Siu (2002) proposed a single-input, single-output (SISO) control scheme based on output measurement and input change after each processing cycle. They also experimentally implemented cycle-to-cycle control of a simple bending process. Rzepniewski and Hardt (2003, 2004) provided the extension of the cycle-to-cycle control concept to the general Multiple-Input Multiple-Output (MIMO) situation. It has been shown that properties of zero mean error and bounded variance amplification that were seen for the SISO case can also be achieved for the MIMO case. Finally, they noted that MIMO cycle-to-cycle control is an appropriate candidate for a system having many thousands of inputs and outputs (e.g., reconfigurable discrete forming die).

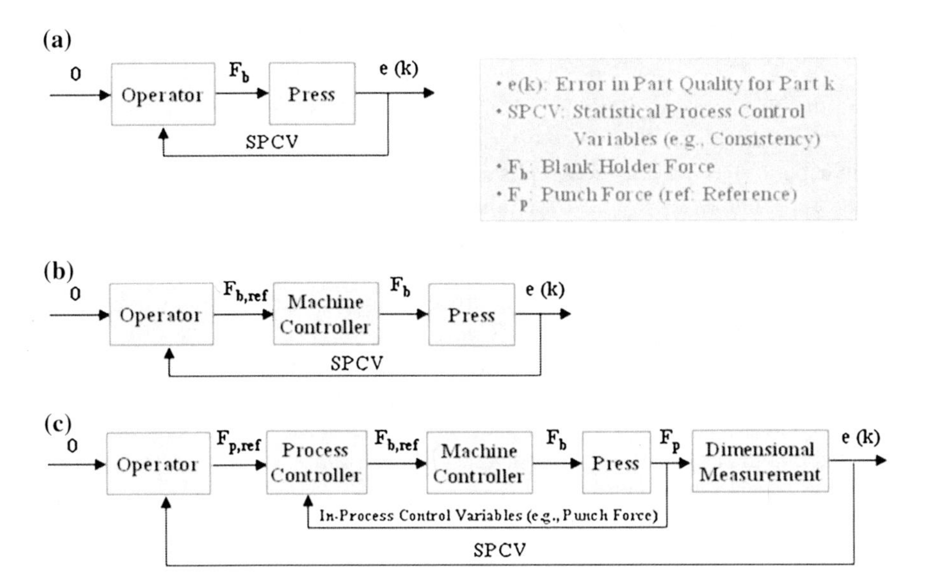

Fig. 3.3 Expected evolution of cycle-to-cycle control, **a** current, **b** with machine control, and **c** with in-process and machine control

Cycle-to-cycle control itself has been used to improve stamped part quality through post-process inspection or in-process variable monitoring. However, post-process corrections can only be achieved after bad parts are produced. Ultimately, in-process control, despite its additional cost and difficulty in sensing, is needed to improve formability, dimensional accuracy, and consistency in production.

3.2 Active Binder Control Systems

In the sheet metal stamping process, the BHF controls the material flow into the die cavity, and optimal material flow plays a critical role in producing a good stamped part. Conventional passive die-cushions filled with nitrogen gas could be replaced with an active BHF control system actuated by multiple hydraulic cylinders (see Sect. 2.5). The objective is to improve the formability and dimensional accuracy of stamped part by varying the BHF at different locations on the die, as well as at different times during the punch stroke.

Recent research shows that die try-out time can be reduced by up to 80 % when an active binder system, controlled by multi-cylinder actuators, is used in die try-out. This is accomplished by varying the BHF at different locations based on punch stroke, instead of grinding and welding.

3.2.1 Segmented Blank Holder System

While the elastic binder (Doege et al. 2001) focused on generating homogeneous binder pressure on the blank of the sheet metal, a segmented blank holder (or flexible binder) is able to accomplish control of one segment while not being significantly influenced by the variation and distribution of other segments.

Yagami et al. (2004) employed segmented blank holder modules to control the material flow into the die cavity, enhancing the effect of the BHF control and improving formability in the stamping process. They obtained fuzzy blank holder pressure (BHP) trajectories for each blank holder segment and showed that the distributed BHP method can improve wall thickness distribution.

Wang et al. (2005) developed a space variant BHF system with segmented blank holders to control the strain path during the deep drawing process. They reported that the key advantage is that strain in the forming process can be adjusted in a safe working area without fracture.

3.2.2 Pulsating BHF Control System

A new approach to the variation of BHF has involved pulsation. Experiments by Ziegler et al. (1999) showed that the onset of wrinkling in a blank drawn with a pulsating BHF occurs at a displacement similar to that obtained under a constant BHF equal to the maximum force of the pulsation (see Fig. 3.4). The reduction in the friction force achieved when the pulse reduces the BHF to below this maximum allows increased deformation to occur prior to tearing, without sacrificing effective wrinkle suppression. An example of the increase in the working window achieved with zinc-coated and phosphated steel sheets, employing a pulse frequency of approximately 3 Hz (the specific frequency itself was determined to be of little influence), is demonstrated in Fig. 3.4. The key objective that he tried to achieve was to avoid cracks on the surface by reducing the friction force. For example, with constant BHF, it was only possible to avoid cracks for the friction coefficient $\mu = 0.1$. With higher friction coefficients cracks occur. With pulsating BHF, it was possible to avoid cracks even if the friction coefficient increased up to $\mu = 0.12$. Ultimately, this showed that pulsating BHF helped to increase the robustness of the process and contributed to avoiding scratches on the surface of stamped parts. However, the amplitude and frequency of the pulses need to be tuned with respect to the lubrication and material properties for a given stamping.

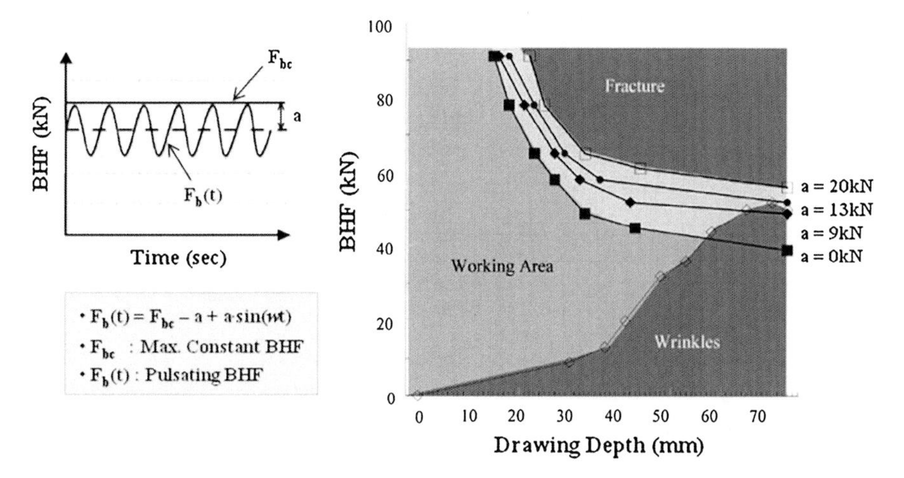

Fig. 3.4 Working area for pulsating BHF with amplitudes of 0 kN (static case), 9, 13, 20 kN; frequency: 3 Hz; sheet material: ZEPH (Ziegler 1999)

3.2.3 Active Drawbead Control System

Material flow in the forming process is often modified locally by the insertion of drawbeads into the tooling. In practice, the drawbead is a fixed component on the dies. However, Michler et al. (1994) and Bohn et al. (1998) implemented their control function with a set of active drawbead actuators. They constructed a multiple-action hydraulic sheet metal strip-drawing tool for the purpose of studying the effectiveness of feedback control in forming. As shown in Fig. 3.5, a punch pulls a strip of sheet metal over a die shoulder and a controllable drawbead is located in the center of the blank holder. Both drawbead penetration and BHF are controlled while the apparatus is measuring and recording the drawbead position, the vertical drawbead force, BHF, and the punch (strip pulling) force (i.e., measured output). In experiments, a PI controller was used, adjusting the drawbead penetration to compensate for the deviation between the reference input (i.e., desired punch force trajectory) and the measured output (i.e., actual punch force).

An active drawbead control system can achieve fast response and require smaller energy consumption than other types of active blank holder systems consisting of large inertia-based hydraulic actuators. However, this idea is difficult to implement in practice, due to complexity and cost in the production of the dies.

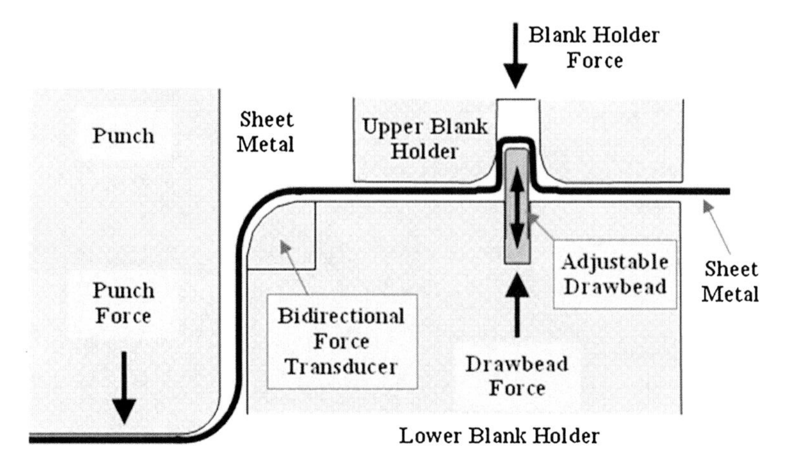

Fig. 3.5 Schematic diagram of the active drawbead control system with bidirectional transducer (Michler et al. 1994)

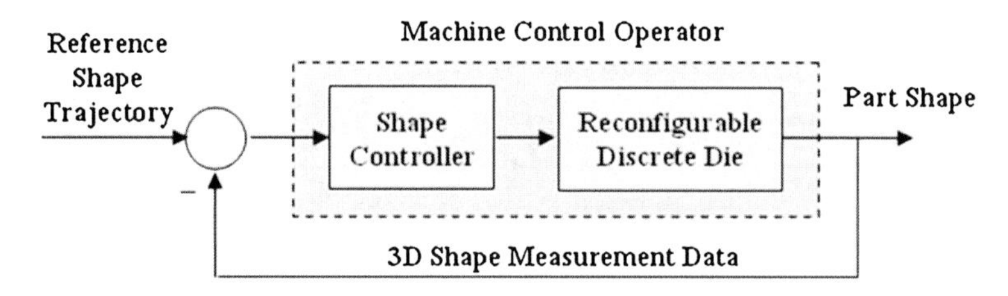

Fig. 3.6 Shape control system using a reconfigurable tool and spatial frequency controller (Hardt 2002)

3.2.4 Reconfigurable Discrete Die

A reconfigurable forming tool attempts to use a die whose shape can be rapidly reprogrammed between forming cycles. If the die surface is in some way programmable, then, the stamped part quality can be improved. Obviously, a key advantage of the reconfigurable die is that it rapidly enables one to regenerate new dies, whose shape is different from previous ones, with aid of die reconfiguration actuators.

Walczyk et al. (1998) and Hardt (2002) addressed the design and analysis issues involved with movable die pins, turning a matrix of die pins into a rigid tool, and the pin matrix containment frame. As illustrated in Fig. 3.6, they proposed a feedback control scheme to monitor directly the 3D shape of the stamped part. Using this approach, the pin actuators are controlled by the shape controller until part shape errors are minimized with respect to a predetermined shape trajectory. The reconfigurable tool was combined with a three-dimensional shape-sensing device and a spatial frequency-based control law. However, the reconfigurable discrete die may not be applicable to produce very complex part shapes. Challenges include optimizing the number of actuator pins with respect to cost and complexity.

3.3 Process Variables and Sensors

There are many opportunities to measure physical quantities either on the machine or the workpiece itself in stamping. Because the most important constitutive relationship for forming is stress-strain or force-displacement, the latter two quantities are most often measured. In general, monitoring process variables (e.g., punch force, draw-in, and wrinkling) in the sheet metal forming process is very important to improve stamped part quality and to reduce cost and time-consuming die-work. Thus, many researchers have focused on sensors to monitor process variables for use in control of the stamping process.

3.3.1 Punch Force

Among the process variables, punch (i.e., strip pulling) force is valuable to interpret the stress-strain curve for the material, because sheet metal pulling force is directly involved in failure (Hosford et al. 1993). The punch force can be measured using commercial sensors installed on the stamping press.

Michler et al. (1994) detected the punch force using a bi-directional force transducer for an adjustable drawbead system that varied drawbead penetration to control the draw-in restraining force. This behavior of the punch force is influenced to a significant extent by the drawbead restraining force.

Similar measurement of the punch force was achieved by Hsu et al. (1999, 2000, 2002) They presented in-process control though adjustment of the BHF using a hydraulically controlled press based on tracking a reference punch force trajectory to improve part quality and consistency.

Sensing the punch force as a process variable in the forming process is easy to implement in practice. However, the measured punch force represents the resultant effect of the forming process and lacks local detail.

3.3.2 Draw-In

The ideal feedback measurement for in-process control of forming would be the stress and strain field throughout the sheet metal. With this information local springback can be reduced and fracture can be also prevented. Unfortunately, in-process measurements of stresses and strains are impractical. However, certain displacements can be measured. In processes where sections of material remain free of surface pressure, mechanical and optical measurement devices could be inserted to sense draw-in of the sheet metal.

Using linear variable differential transducers (LVDTs), Hardt et al. (1993) measured draw-into control the BHF in process to ensure optimal forming conditions. Then, they proved that displacement of the edge of the sheet during draw-in was not reliable because of tearing. They also proposed a method that measured the circumferential contraction of the material, in averaging all draw-in over the entire circumference of the blank.

Sunseri et al. (1996) and Siegert et al. (1997) also used an LVDT type draw-in sensor to reduce springback and wrinkling respectively. However, the LVDT requires significant setup time in practice, and becomes too time-consuming and expensive to use in production.

A compact, economic draw-in sensor to overcome the week-point of the LVDT type sensor has been developed. Lo et al. (1999) monitored the displacement of sheet metal blank, using a reflective photoelectric encoder, which has a rotating wheel where it contacts the sheet metal. However, this sensor can detect only one direction as the sheet metal moves tangentially with respect to the rotating wheel.

Doege et al. (2002) developed a computer-mouse-like, ball sensor, which is based on the mechanical transmission of the plane movement of the sheet metal onto a ball. Using the ball sensor the material draw-in direction, material flow velocity and material flow path can be independently measured in two orthogonal directions. Doege et al. (2003) also designed a computer-mouse-like, contactless-optical sensor for online sheet metal flow measurement. This contactless-optical sensor consists of a chip in which a complementary metal oxide semiconductor (CMOS) sensor and a digital signal processor (DSP) are integrated. One point from the sheet surface is analyzed by the sensor and described at its initial position by the two pixel values (i.e., PX1 and PY1). When the object is moved, the image point moves to a different position with the pixel values (i.e., PX2 and PY2). Sensing accuracy of optical sensors in draw-in displacement was improved at each local position, compared to the contact-based draw-in sensor. However, on the lower stamping die there are still implementation difficulties and cost challenges for practical use.

Cao et al. (2001, 2002) developed a new type of draw-in sensor, which has two key advantages: ease of setup and cost-effective implementation in industrial applications. The installation of LVDT and optical sensors requires either setup time with each forming cycle or intricate tooling modification. Based on the mutual inductance principle, they designed a draw-in sensor by experimenting with a prototype printed on a conventional circuit board to address the need for an affordable and accurate draw-in sensor. This design of sensor was small enough to be embedded in a die or blank holder. In the single transducer configuration the primary and secondary coils, as shown in Fig. 3.7a, were printed into one transducer board. Utilizing the principle of mutual inductance between the two loops, the linear draw-in of sheet metal was detected based on the uncovered area of the primary and secondary coils on the board, as seen in Fig. 3.7b. The linear position

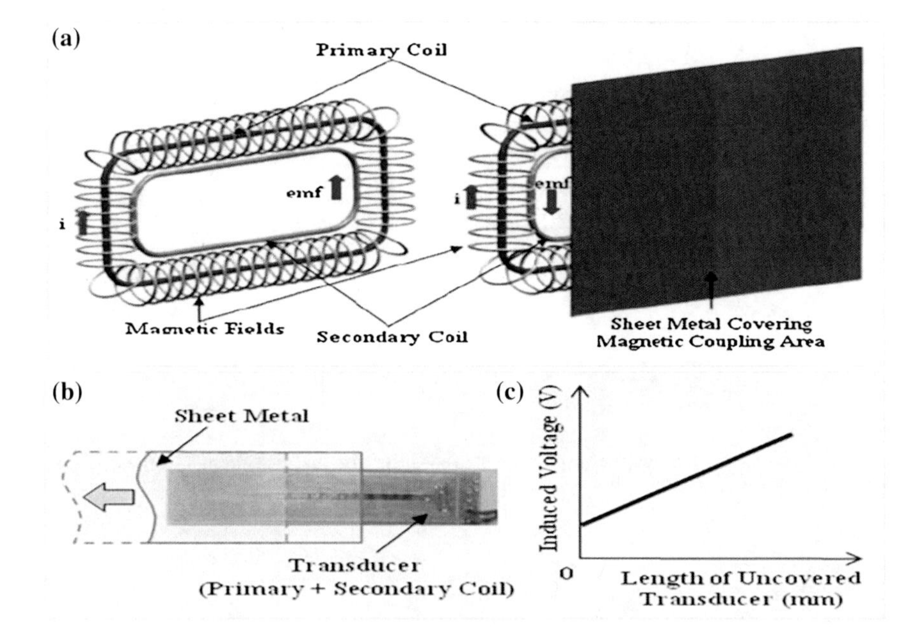

Fig. 3.7 **a** Operating principle of transducer, **b** transducer and sheet metal configuration, **c** Induced voltage signal from transducer (Cao et al. 2002; Mahayotsanun et al. 2005)

sensor transmitted signals to a signal conditioning board, which amplified and filtered the induced voltage readings and these readings were sampled using a computer based data acquisition system. Thus, sheet metal draw-in can be obtained using the voltages generated by the draw-in sensor, after calibration using an LVDT, as illustrated in Fig. 3.7c. However, the sensor has to be calibrated for each material used and the inductive characteristics are dependent on material properties. Consequently, if this sensor is able to demonstrate endurance for a large number of stamping cycles, it may become adopted by industry due to its ease of use and low cost. In particular, thick epoxy (i.e., about $0.8 \sim 2$ mm) covers the top of the sensor to protect it from wearing out over many hours operation and to place constant gap between the transducer and the sheet metal. However, it may have errors due to wrinkling, which creates varying gap conditions between the transducer and the blank.

3.3.3 Wrinkling

The wrinkling of sheet metal is a common phenomenon, which arises in forming due to compressive stresses. The ability to sense the occurrence of wrinkles is

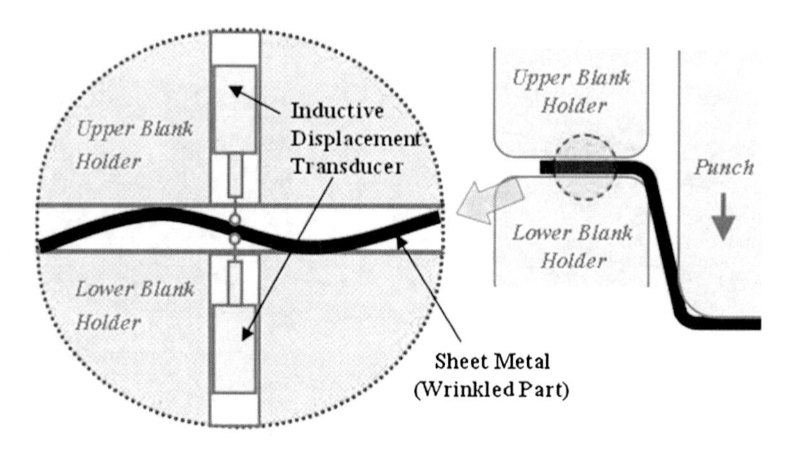

Fig. 3.8 Schematic of inductive displacement transducer for the measurement of the wrinkle height (Siegert et al. 1997)

potentially useful in the sheet metal forming process consisting of closed-loop process control systems (e.g., active BHF).

Pereira et al. (1994) presented a method using two fiber optic displacement sensors for detecting low and high frequency wrinkling in stamping. From two parallel non-contact readings attached to the upper binder, estimation of the peak amplitude of the wrinkle was achieved by combining estimation of wrinkle frequency (ω) with the distance between two sensors. Though this non-contact based optical wrinkle sensor would be free from the wear problem in applications, it has also a technical challenge; it is difficult to choose the optimal distance between two readings based on the smallest wrinkling frequency in order to avoid aliasing (e.g., two or more oscillations of wrinkling within the distance between two readings).

The measurement of wrinkle height, as shown in Fig. 3.8, was achieved in closed-loop stamping by applying a combination of two opposing displacement transducers, which are positioned in the upper binder and the lower binder (Siegert et al. 1997). The displacements of the two transducers can be used to measure the real wrinkle height. Changes in sheet thickness cause errors in the measurement of the height if only the displacement between upper and lower binder is measured. However, this contact-based wrinkling sensor is limited in industrial application because of friction-based endurance failures at the sensor tip that contacts the sheet, and also because wrinkle locations are not known a priori.

3.4 Concluding Remarks

This chapter reviews key research developments in feedback control of the sheet metal stamping process and its effect on the quality of stamped parts. The use of feedback control to improve part quality requires addressing several technical issues, including the generation of accurate reference trajectories for the control loops using FEM or design-of-experiments. In-process control also requires the implementation of controllers to adjust BHF and achieves the control objective of tracking the desired reference trajectories. The design of these controllers requires accurate models of the forming process. In addition, the development of reliable, cost-effective sensors to measure representative process variables is also a key technical challenge. Addressing these issues will lead to the creation of systems that combine statistical process control methods, machine control, in-process control, and cycle-to-cycle control capabilities to significantly improve part quality and consistency in the stamping process.

References

Adamson, A., Ulsoy. A.G., Demeri, M. (1996). Dimensional Control in Sheet Metal Forming via Active Binder Force Adjustment. SME Transactions. Vol. 24 167–178.

Ahmetoglu, M.A., Altan, T., Kinzel, G.L. (1992). Improvement of Part Quality in Stamping by Controlling Blank-Holder Force and Pressure. J. of Materials Processing Tech. Vol. 33 195–214.

Bohn, M. L., Jurthe, Stefen U., Weinmann, Klaus J. (1998). A New Multi-point Active Drawbead Forming Die: Model Development for Process Optimization. SAE Paper. No. 980076 24–30.

Cao, Jian., Kinsey, B.L., Yao, Hong., Viswanathan, Vikram., Song, Nan. (2001). Next Generation Stamping Die – Controllability and Flexibility. Robotics and Computer Integrated Manufacturing. Vol. 17 49–56.

Cao, Jian., Lee, J. Peshkin, M. (2002). Real-time Draw-in Sensors and Methods of Fabrication. Northwestern University U.S. Pat. 6,769,280.

Doege, E., Elend, L.-E. (2001). Design and Application of Pliable Blank Holder Systems for the Optimization of Process Conditions in Sheet Metal Forming. J. of Materials Proc. Tech. Vol. 111 182–187.

Doege, E., Seidel, H.-J., Griesbach, B., Yun, J.-W. (2002). Contactless on-line Measurement of Material Flow for Closed Loop Control of Deep Drawing. J. of Materials Proc. Tech. Vol. 130–131 95–99.

Doege, E., Schmidt-Jurgensen, R. Huinink, S., Yun, J. W. (2003). Development of an Optical Sensor for the Measurement of the Material Flow in Deep Drawing Processes. CIRP Annals – M. Tech. Vol. 52 225–228.

Hardt, D. (2002). Forming Processes: Monitoring and Control. Mechanical Systems Design Handbook. (O. Nwokah and Y. Hurmuzlu (Ed.)), 105–119. CRC Press

Hardt, D. (1993). Modeling and Control of Manufacturing Processes: Getting More Involved. J. of Dynamic Systems, Meas. and Control. Vol. 115 291–300.

Hardt, D., Fenn, R. C., (1993). Real-Time Control of Sheet Stability during Forming. In: Trans. ASME, J. Eng. Ind., Vol. 115 301–308.

Hardt, D., Siu, T.S. (2002). Cycle-to-Cycle Manufacturing Process Control. 1st Annual SMA Symposium, Singapore

Herderich, M.R. (1990). Experimental Determination of the Blankholder Forces Needed for Stretch Draw Die Design. SAE Paper. No. 900281 53–61.

Hosford, W.F., Caddell, R. M. (1993). Metal Forming: Mechanics and Metallurgy. 286–308. PTR Prentice-Hall. 2nd Edition.

Hsu, C.W., Ulsoy, A.G., Demeri, M.Y. (2002). Development of Process Control in Sheet Metal Forming. J. of Mater. Proc. Tech. Vol. 127 717–724.

Hsu, C.W., Ulsoy, A.G., Demeri, M.Y. (2000). An Approach for Modeling Sheet Metal Forming for Process Controller Design. ASME J. Manuf. Sci. Eng. Vol. 122 717–724.

Hsu, C.W., Ulsoy, A.G., Demeri, M.Y. (1999). Process Controller Design for Sheet Metal Forming. American Control Conference. Vol. 1 192–196.

Kergen, R., Jodogne, P. (1992). Computerized Control of the Blankholder Pressure on Deep Drawing Process. SAE Paper. No. 920433 51–55.

Kinsey, B., Cao, J. Solla, S. (2000). Consistent and Minimal Springback Using a Stepped Binder Force Trajectory and Neural Network Control, J. Eng. Mater. Technol., 122, pp. 113–118.

Lo, Sy-Wei., Jeng, Guo-Ming. (1999). Monitoring the Displacement of a Blank in a Deep Drawing Process by Using a New Embedded-Type Sensor. Int. J. Adv. Manuf. Technol. Vol. 15 815–821.

Mahayotsanun, N., Cao, Jian, Peshkin, Michael, (2005). A Draw-In Sensor for Process Control and Optimization. American Institute of Physics, CP778 Vol. A.

Manabe, K., Koyama, H., Katoh, K., Yoshihara, S. (1999). Intelligent Design Architecture for Process Control of Deep-Drawing. Intel. Proc. and Manufacturing of Materials. Vol. 1 571–576.

Michler, J.R., Weinmann, K.J., Kashani, A.R., Majlessi, S.A. (1994). A Strip Drawing Simulator with Computer-Controlled Drawbead Penetration and Blank Holder Pressure. J. Materials Proc. Tech. Vol. 43 177–194.

Pereira, Pratap., Zheng, Y.F. (1994). Sensing Strategy to Detect Wrinkles in Components. IEEE Transactions on Instrum. and Measurement. Vol. 43 No. 3 442-447.

Rzepniewski, Adam K., Hardt, David. E. (2004). Multi-input Multi-Output Cycle-to-Cycle Control of Manufacturing Processes. 3rd Annual SMA Sympo.

Rzepniewski, Adam K., Hardt, David. E. (2003). Gaussian Distribution Approximation for Localized Effects of Input Parameters. 2nd Annual SMA Symposium

Sheng, Z. G., Jirathearnat, S., Altan, T. (2004). Adaptive FEM Simulation for Prediction of Variable Blank Holder Force in Conical Cup Drawing. International J. of Machine Tools & Manuf. Vol. 44 487–494.

Siegert, K., Hohnhaus, J., Wagner, S. (1992). Combination of Hydraulic Multipoint Cushion System and Segment-Elastic Blankholders. SAE Paper. No. 98007 51–55.

Siegert, K., Ziegler, M., Wagner, S. (1997). Loop Control of the Friction Force: Deep drawing process. J. of Materials Proc. Tech. Vol. 71 126–133.

Sklad, M.P., Harris, C.B., Slekirk, J.F., Grieshaber, D.J. (1991). Modeling of Die Tryout. SAE Paper No. 920433 151–157.

Sunseri, M., Cao, J., Karafillis, A.P., Boyce, M.C. (1996). Accommodation of Springback Error in Channel Forming Using Active Binder Force: Control Numerical Simulations and Experiments. J. of Engin. Materials and Tech. Vol. 118 426–435.

Viswanathan, V., Kinsey, B., Cao, J. (2003). Experimental Implementation of Neural Network Springback Control for Sheet Metal Forming. J. of Engin. Materials and Tech. Vol. 125 141–147.

Walczyk, D.F., Hardt, D.E. (1998). Design and Analysis of Reconfigurable Discrete Dies for Sheet Metal Forming. J. Manuf. Systems. Vol. 17 No. 6 436–454.

Wang, Lin., Lee, T.C. (2005). Controlled strain path forming process with space variant blank holder force using RSM method. J. of Materials Processing Technology. Vol. 167 447–455.

Xu, Siguang., Zhao, Kunmin., Lanker, Terry., Zhang, Jimmy., Wang, C. T. (2007). On Improving the Accuracy of Springback Prediction and Die Compensation. SAE Paper No. 2007-01-1687

Yagami, T., Manabe, Ken-ichi, Yang, M., Koyama, H. (2004). Intelligent Sheet Stamping Process Using Segment Blankholder Modules. J. of Materials Proc. Tech. Vol. 155–156 2099–2105.

Zhong-qin, Lin., Wang, Wu-rong, Chen, Guan-long. (2007). A New Strategy to Optimize Variable Blank Holder Force towards Improving the Forming Limits of Aluminum Sheet Metal Forming. J. of Materials Processing Technology. Vol. 183 339–346.

Ziegler, M. (1999). Pulsating Blankholder Technology. SAE Paper. No. 1999-01-3155 1–5.

Chapter 4
Machine Control

Abstract Stamping process control is achieved by commanding actuation mechanisms to appropriately vary material flow into the die during the stamping cycle, based on feedback on part quality. The actuation mechanisms in turn employ closed-loop control to achieve the desired material flow. Thus, process control forms an outer loop while control of the actuators form an inner loop. The inner-loop control is referred to as machine control and in this chapter, the non-linear dynamics of hydraulic actuators for variable binder-force control are analyzed and machine control approaches for these actuators are described. Real-time digital implementation of machine control algorithms is also considered.

4.1 Machine Control Objectives

Referring to Sect. 2.5 and Fig. 2.8, it can be seen that the machine controller ensures that the material flow actuators deliver the desired restraining forces on the blank, as commanded by the process controller. As the material flow needs to be controlled locally, multiple actuators are required, with each actuator acting independently.

The discussion in this book is limited to hydraulic actuation methods for variable binder force control, which are the only forms of material flow control that have been implemented in production stamping presses to date. These methods involve the use of multiple hydraulic cylinders mounted under the lower binder (see Fig. 1.6) or in the press-bed in the form of a press cushion.

The machine control objective for a hydraulic variable binder force system is to ensure that the force delivered by the pressurized fluid in each hydraulic cylinder tracks the reference force command generated for that cylinder to maintain consistent part quality. The reference force command for each cylinder can be time-varying during the cycle; it is typically constructed as the sum of a baseline command obtained during die try-out and the output of the process controller.

Y. Lim et al., *Process Control for Sheet-Metal Stamping*,
Advances in Industrial Control, DOI: 10.1007/978-1-4471-6284-1_4,

The machine controller also has to reject disturbances and be robust to variations such as changes in the bulk-modulus in the hydraulic fluid due to temperature variation. The step-response should be sufficiently fast to ensure accurate material flow control, with a settling time of about 0.05 s, given that a typical stamping cycle lasts between 0.5 and 1 s.

In the next section, the working principle and construction of hydraulic actuation for variable binder force systems will be described and the dynamic equations of the actuators will be derived, showing the nonlinearities that need to be addressed during controller design.

4.2 Design of Variable Binder Force Systems

A typical hydraulic actuation system consists of a set of hydraulic cylinders mounted under the lower binder in a press. The cylinders are single acting, that is, they are only filled with fluid on the piston-side. The binder force in each cylinder is created by the compression of the hydraulic fluid in it by the downward motion of the binder as it moves with the press ram. The fluid in each cylinder is metered out to the tank using a dedicated servo-valve, allowing each cylinder to be controlled independently. The pressure in each cylinder is measured using a standard industrial capacitive or piezo pressure sensor, and the force created by the cylinder can be estimated by multiplying by the cross-sectional area of the cylinder piston. A displacement sensor that measures the travel of the binder during the stamping cycle is also necessary as it is used for two purposes. The first is to ensure that the appropriate force is commanded at each point in the stamping cycle. The baseline reference force profile is matched to displacement from the bottom of the cycle during try-out. The second is to implement the machine control law given by (4.2) and (4.4).

After the stamping cycle is complete and the press ram moves up and out of contact with the lower binder, the lower binder needs to be raised by extending the cylinders. This can be done by using a pump to drive hydraulic fluid back into the cylinders; however, to meet the cycle times in typical production presses with a large number of cylinders, a large and expensive pump may be required. A simpler, less expensive and more reliable method is to use a tank pressurized using compressed shop-air. Figure 4.1 shows a schematic showing the components for a typical hydraulically actuated variable binder-force system. We now analyze the equations that represent the hydraulic dynamics in each cylinder as it is driven down by the press ram. Using the approach and notation in Sect. 2.3, the rate of change of pressure in the cylinder during the forming stroke is given by

$$\dot{P}_i(t) = \beta \frac{A_c\dot{d}(t) - \alpha_i(t)K_v\sqrt{\frac{2(P_i(t)-P_T)}{\rho}}}{A_c(l_c - d(t))} \tag{4.1}$$

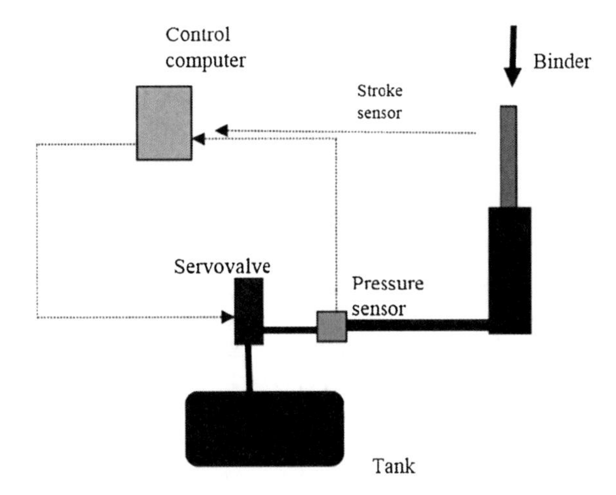

Fig. 4.1 Hydraulic actuation schematic for variable binder force system

where $P_i(t)$ denotes the pressure in the ith cylinder at time instant t, A_c denotes the cross-sectional area of the cylinder on the piston-side, l_c denotes the length of the hydraulic fluid column in the cylinder, $d(t)$ is the travel of the cylinder piston at time instant t, $0 \leq \alpha_i(t) \leq 1$ is the valve-opening ratio with 0 denoting the valve fully closed and 1 denoting it fully open, and P_T is the tank pressure. $\alpha_i(t)$ is the control input used to create the desired pressure profile in the cylinder, and it assumed to be proportional to the command voltage applied to the servo-valve. It is also noted that $\dot{d}(t)$ can be assumed to be the same as the press ram speed, $\dot{y}(t)$, after the ram comes in contact with the binder, and thus the rod of the cylinder which supports the binder.

Clearly (4.1) is nonlinear and it can be seen that as the cylinder comes close to the bottom of the stroke, $l_c - d(t)$ is small and thus, $\dot{P}_i(t)$ is amplified; this makes precise control at the bottom of the forming cycle, which is also the most crucial phase of the cycle, extremely important.

The critical factors to be considered in the design of the actuation system are the desired maximum force (tonnage) that is to be developed in each cylinder and the maximum rate of travel of the binder. The total binder force typically required for large automotive parts is in the range of 150–300 tons. Using cylinders with a 10 cm bore and assuming a maximum operating pressure of 32 MPa, each cylinder will deliver approximately 25 tons. Thus, a variable binder force system with 12 such cylinders will be able to provide 300 tons of binder force which should be sufficient to form parts even from high-strength steel. For most parts, the cylinder should be designed to allow about 20–25 cm of travel.

It can be seen from (4.1) that to maintain a constant pressure the flow rate of fluid out of the cylinder through the servo-valve into the tank should be $A_c\dot{y}(t)$. Thus, the hoses should be sized to carry a flow-rate corresponding to the maximum value of $\dot{y}(t)$, which is the speed of the binder as the die closes. For cylinders with dimensions such as those described in the previous paragraph, operating with a maximum binder speed of 35 cm/s, 2 cm diameter ultra-flexible hoses are recommended, considering that flexibility is very important in being able to route the hoses under the binder. These hoses need to be rated for operation of 35 MPa. An example of a commercially available hose that meets these specifications is the Parker 787TC constant working pressure hose.

The next major component in the design is the servo-valve. The two main design parameters for choosing an appropriate servo-valve are speed of response and flow-rate. As the typical stamping cycle lasts anywhere between 0.5 and 1 s, and it may be necessary to make up to 10 binder force changes in this period, it can be seen that it would be desirable to have a valve with a step-response settling time of about 25 ms. Furthermore, for the operating conditions mentioned above, flow-rates of up to 160 l/min can be expected and thus, the servo-valves need to be sized to accommodate such flows, with minimal pressure build-up. The Moog D634 servo-proportional valve and the Woodward R-DDV 27G series servo-valve are examples of commercially available valves that provide the required performance, when configured for double-ported (2×2 way) operation.

Standard capacitive or piezoelectric pressure sensors with a range of up to 35 MPa can be used to measure the pressure in each cylinder from which the force can be derived and LVDTs or other displacement sensors can be used to measure the travel of the binder.

Finally, to ensure that the system operates safely, pressure relief valves need to be connected in parallel with the servo-valves on the same hydraulic lines that run to the cylinders.

The pressure tanks need to be designed to withstand the pressure surges that will occur when hydraulic fluid rushes out of the cylinders into the tank. As the fluid can heat up during operation, copper chilled-water cooling tubes can be mounted in the tanks. The thermal design requirements are dependent on the cycle rate of the press during operation and need to be assessed for each particular use case.

4.3 Controller Design

The simplest approach to designing a machine controller would be to close a simple PID loop using pressure feedback. However, as noted in the previous section, the hydraulic dynamics are nonlinear, and in the case of a mechanical press, $\dot{d}(t)$ is high at top of the stroke and drops to zero at the bottom of the stroke. Thus, high gain is required initially, but near the bottom of the stroke, where splitting is most likely to occur, high gain can cause instability, leading to undesirable binder forces and poor part quality.

An alternative design approach is one that uses *feedback linearization* to address the nonlinear dynamics of the system (Isidori and Kerner 1982). Feedback linearization is a technique by which a nonlinear system is effectively linearized by canceling out the nonlinear terms in the plant dynamics by introducing additional terms in the controller. While from a theoretical perspective, it can be viewed as a non-robust control method because canceling out the nonlinear terms requires precise knowledge of these terms, and little stability analysis has been performed on the effect of nonlinear residuals, this method has been very successfully used in several practical hydraulic applications; see, for example, Seo et al. (2007).

For hydraulic actuation of variable binder force systems, a feedback linearizing controller can easily be obtained by choosing

$$\alpha_i(t) = \frac{1}{K_v}\sqrt{\frac{\rho}{2(P_i(t) - P_T)}}\left(A_c\dot{d}(t) - \frac{\vartheta(t)A_c(l_c - d(t))}{\beta}\right) \tag{4.2}$$

where $\vartheta(t)$ is an additional control variable. Substituting, (4.2) in (4.1), it follows that

$$\dot{P}_i(t) = \vartheta(t) \tag{4.3}$$

and, thus, we have now used feedback linearization to transform the system into first-order linear system with control input $\vartheta(t)$. A PI loop using pressure feedback can now be introduced with the control input $\vartheta(t)$, that is, by choosing

$$\vartheta(t) = K_P\big(P_{i,ref}(t) - P_i(t)\big) + \int K_i\big(P_{i,ref}(t) - P_i(t)\big)dt \tag{4.4}$$

Figure 4.2 shows simulation results using both simple PI control and feedback-linearized PI control. The simulation assumes that the press ram of a mechanical press impacts the cylinder rod at 0.5 m/s. The press parameters are assumed to be those in Sect. 2.2, except that the press is assumed to be running at 12 strokes/min, a speed more typical of presses running larger parts for which variable binder force is typically used. The cylinder piston diameter (or bore) is taken to be 10 cm ($A_c = 0.0081$ m^2), the travel is taken to be 20 cm ($0 \le d(t) \le 0.2$ m with $l = 25$ cm. The servo-valve is assumed to have a flow-rate of 0.005 m^3 at a pressure differential of 6.9 MPa and a step-response settling time of 25 ms.

The desired binder force profile, shown as the red dotted-line, is typical with a low-to-moderate tonnage of 8 tons at the beginning of the stroke to set the drawbead. The tonnage is then ramped up to 20 tons during the initial part of the forming process but is then dropped to 4 tons avoid splits in the latter part of the cycle. At the end, the tonnage is ramped back up to 8 tons to avoid springback. It can be seen that both simple PI (green line) and the feedback linearized controller (blue line) track the reference profile well until the last 1 cm (or 0.1 s) of the forming cycle. (Note that the top plot goes from right to left as the horizontal axis represents the distance from the bottom of the stroke). However, at the end of the cycle, the simple PI controller shows poor tracking performance and this can be crucial to part quality. However, the feedback-linearized controller provides very high tracking performance throughout the forming cycle.

Figure 4.3 shows the servo-valve opening factor $\alpha_i(t)$ for this simulation.

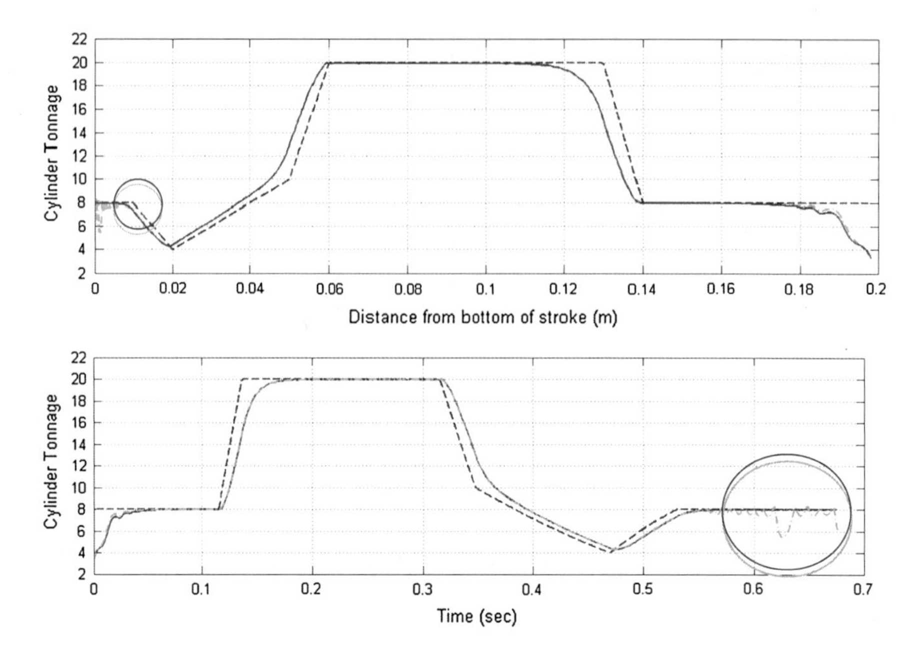

Fig. 4.2 Hydraulic cylinder force (tonnage) using simple PI (*green*) and feedback-linearized PI (*blue*) control. *Top* plot is with respect to distance from *bottom* of stroke; *bottom* plot is with respect to time

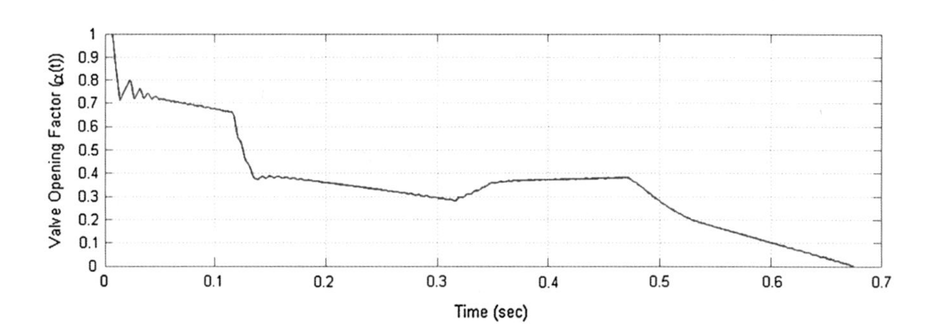

Fig. 4.3 Valve opening factor using feedback-linearized PI control

The valve is initially fully opened to compensate for the fast motion of the press ram at the beginning of the stroke. Then, the valve continues to close during most of the cycle to maintain the desired pressure in the cylinder. This is due to the fact that the press ram speed is decreasing.

The feedback linearized controller presented in this section can be used for both hydraulic and mechanical presses; however, it requires an estimate (or measurement) of the cylinder rod (or press ram) speed. If it is estimated from the displacement measurement of the binder, a low-pass filter on the measurement is advised as differentiation amplifies noise.

It is also noted that a single displacement sensor can be used to measure the rod displacements and speeds of all the cylinders in the system as the binder flexure is of a much smaller scale in comparison to the motion of the cylinder rods.

A discrete-time version of the feedback linearized control algorithm described above is detailed in Venugopal (2009). It also includes a method for adjusting the tonnage reference command with respect to the press-speed to enable better tracking performance.

4.4 Controller Implementation

Controller implementation will be discussed considering two aspects:

1. The realization of the controller on real-time hardware;
2. The development of a shop-floor appropriate interface.

First, we note that the machine controller in (4.2) and (4.4) has to be implemented using a real-time computer, as standard analog PID boards or PLC systems are insufficient. The bandwidth of the servo-valve dictates the sample rate at which the controller is run; the typical rule of thumb is to run at 10 times the frequency of the fastest dynamics of the system. Experimental implementations have shown that a sampling frequency of 1 kHz is sufficient.

The continuous time controller equations need to be numerically integrated with a time-step corresponding to the sampling rate of the system. Simple Euler integration has been shown to have sufficient numerical accuracy; however, second or higher order Runge–Kutta methods can also be easily implemented given the computational capability of today's processors.

The easiest way to implement the control structure described in Sect. 4.3 is to use rapid control prototyping tools in which the controller is constructed in a model-based simulation environment (MBSE) tool such as Simulink® of The Mathworks or LabView from National Instruments and use their code-generation tools to generate real-time code. These tools not only generate code for the control algorithm, but also for real-time scheduling, reading of pressure and displacement sensor data via analog-to-digital (A/D) converters and the generation of servo-valve command voltages via digital-to-analog (D/A) converters. These software tools are closely integrated with several real-time hardware platforms such as those provided by OPAL-RT Technologies, dSPACE and National Instruments. The biggest advantage of these tools is that the control designer does not need to be a skilled real-time software developer and the control algorithm is represented in an easy-to-understand block-diagram form rather than as several pages as code.

The hardware platform to be chosen needs to be ruggedized for operation in a shop-floor environment and be as compact as possible. The PC-104 standard provides a good non-proprietary real-time hardware solution that is well suited for variable binder force systems. The standard is related to a compact form-factor and several companies such as RTD Embedded Technologies and Diamond Systems manufacture processor boards as well A/D and D/A input and output boards conforming to the PC-104 standard. 16-bit resolution or higher is advised for the A/Ds and D/As given the electrical nose levels in shop-floor environments.

The combination of MBSE tools integrated with commercially available real-time hardware also allows for the implementation of the more complex process control algorithms that will described in the following sections. Using these tools, the process and machine controllers can be implemented on the same platform and integration issues can be avoided.

The next part of the discussion on implementation turns to the user interface for the system. The controller software must allow the user to perform the following functions:

1. Input the desired binder force trajectory for each cylinder when the system is being used for try-out with only machine control;
2. Save the history of changes in binder force trajectories;
3. Determine if there is a fault with the sensors, actuators and control processing unit;
4. Integrate process control when try-out is complete and production begins.

The typical operator will not be well-versed in real-time digital systems and thus, the user interface has to be simple and intuitive. Most real-time system vendors provide an application program interface to transfer data to the control processor. A column of tonnage values for each cylinder in increments of 1 cm as part of a data sheet accomplishes the first objective. The ability to import data from FEA analysis is also required. The second objective is achieved by saving the history of this data sheet as try-out progresses along with a text field to record part quality with that particular binder force configuration. Providing this feature is important as try-out is often done using trial-and-error and it is important to be able to go back to best setting if certain changes result in degraded part quality. Monitoring of system functioning is very important as loss of binder-force can lead to excessive wrinkling which can cause irreversible damage to the die surface. On the other hand, excessively high binder force can stall and damage the press. Finally, the system should allow the operator to save the variable binder force trajectories for the best part quality under nominal conditions and also record the profiles of the process variables under those operating conditions. The operator should also be able to match sets of specific actuators (cylinders) to specific process control feedback sensors. The integration of process control with machine control will be discussed in Chaps. 5, 6, 7, 8.

Fig. 4.4 Overview of hydraulic variable binder force system installed for operation in a mechanical press

Fig. 4.5 Cylinders and displacement sensor for hydraulic variable binder force system

4.5 Experimental Results

In this section, experimental results from a variable binder force system deployment in a commercial try-out and low-volume production shop is detailed. Figure 4.4 shows the overall layout of the system. It is portable and can be moved between presses and used for different dies Fig. 4.5 shows a close-up view of the cylinders mounted under the lower binder with the displacement sensor for measuring $d(t)$. The control unit is based on PC-104 hardware supplied by RTD Embedded Systems. The operator user interface is shown in Fig. 4.6.

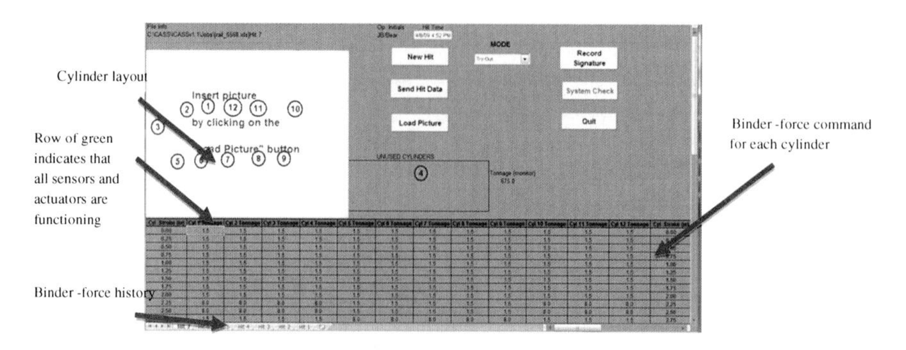

Fig. 4.6 Machine control operator user interface

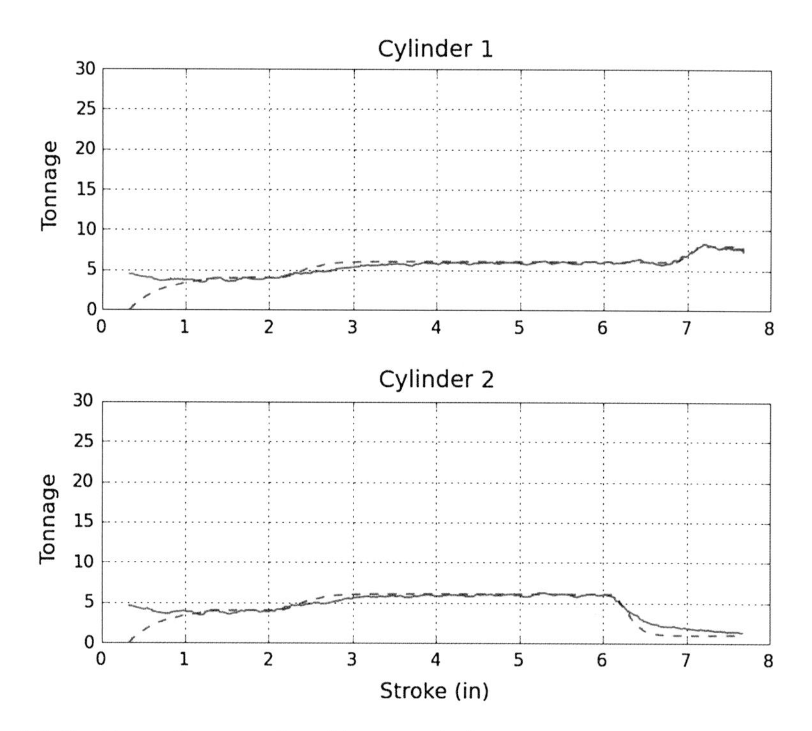

Fig. 4.7 Tracking performance of machine controller in a try-out press

The control algorithm implemented is the one described in Venugopal (2009). Figure 4.7 shows the machine controller tracking performance in a mechanical try-out press that runs at 14 strokes per minute. The tonnage commands, shown as the red-dotted lines, are at relatively low force levels, with a force of about 6 tons for most of the stroke. At the end of the stroke, the tonnage is increased to about 8 tons in Cylinder 1 and dropped to about 2 tons in Cylinder 2. The actual forces delivered by each cylinder (blue lines) closely track the commands.

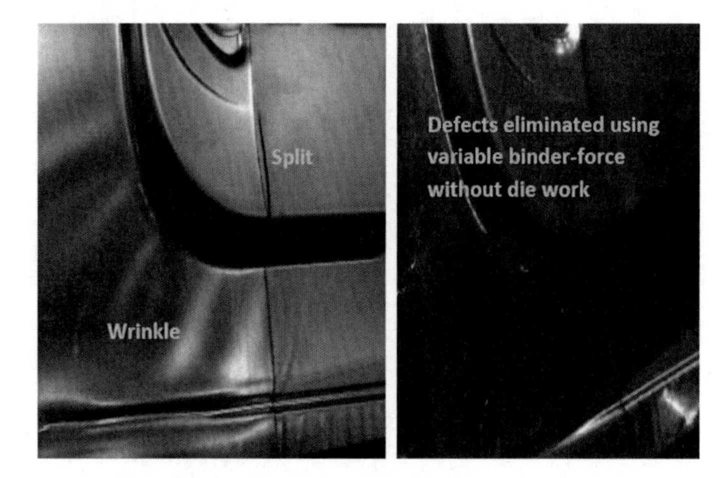

Fig. 4.8 Forming defects eliminated using variable binder force

To close this chapter, we show a defect that is difficult to address by conventional means, corrected using variable binder force under machine control. In Fig. 4.8, we see two defects in the picture on the left, namely a split and a wrinkle in close proximity. These defects are common in the tailor-welded blank used for the part. On the picture on the right, it can be seen that the defects have been eliminated by using variable binder force and increasing the tonnage of the cylinders on both sides of the split but dropping the tonnage in the cylinder to the right of the split about 1 cm from the end of the stroke. The reduced tonnage in one cylinder allows the split near the weld-line to heal while eliminating the wrinkles around it.

References

Isidori, A., Kerner, A. J. (1982) On the feedback equivalence of nonlinear systems. Systems and Control Letters 2: 118–121.

Seo, J., Venugopal, R., Kenné, J. P. (2007). Feedback linearization based control of a rotational hydraulic drive. Control Engineering Practice, 15(12): 1495–1507

Venugopal, R. (2009) Method and system for achieving force control in externally driven hydraulic cylinders, U. S. Patent 7,493,194 B2.

Chapter 5
Laboratory Development of Process Control

Abstract In sheet metal forming processes the blank holder force controls the material flow into the die cavity, which is critical to producing a good part. Process control can be used to adjust the blank holder force in-process based on tracking a reference punch force trajectory to improve part quality and consistency. Key issues in process control include process modeling as well as process controller and reference punch force trajectory design. In this chapter a systematic approach to the design and implementation of a suitable process controller and an optimal reference punch force trajectory is presented. The approach includes the modeling for controller design of the sheet metal forming process, design of the process controller, and determination of the optimal punch force trajectory. Experimental results from U-channel forming on a laboratory forming simulator show that a suitable process controller can be designed through simulation and an optimal reference punch force trajectory can be synthesized through experiments. The proposed development will be useful in designing and implementing process control in sheet metal forming processes as described in subsequent chapters.

5.1 Background on Process Control for Stamping

The control of material flow into the die cavity is crucial for good part quality and consistency, and the blank holder is used to control the material flow. Previous research has shown that varying the blank holder force during forming can improve part quality (Adamson et al. 1996; Ahmetoglu et al. 1995; Schmoekel and Beth 1993) and consistency (Adamson et al. 1996; Hsu et al. 1999). It is worth pointing out that mechanical presses are being retrofitted with hydraulic multi-point cushion systems to provide more control of the forming process (Siegert et al. 1998; Lim et al. 2010, 2012) and many new stamping presses are hydraulic in design. Such press technologies will facilitate the implementation of the process control ideas presented in this book.

Y. Lim et al., *Process Control for Sheet-Metal Stamping*,
Advances in Industrial Control, DOI: 10.1007/978-1-4471-6284-1_5,

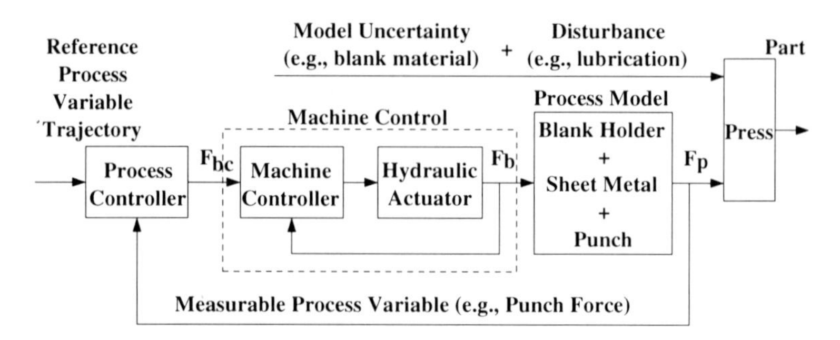

Fig. 5.1 Process control of sheet metal forming

As discussed previously, a strategy for controlling sheet metal forming through the application of variable blank holder force is process control (see Fig. 5.1). In this strategy a measurable process variable (e.g., punch force) is controlled by following a predetermined (e.g., punch force vs. punch stroke or vs. time) reference trajectory through manipulation of the blank holder force (Adamson et al. 1996; Hsu et al. 2002). This strategy was able to produce cups with "optimal" height regardless of initial blank holder force and friction conditions (Hardt and Fenn 1993). Other measurable process variables (e.g., draw-in and friction force) have also been reported (Siegert et al. 1995, 1997; Sim and Boyce 1992).

To systematically design a suitable process controller, the process model in Fig. 5.1 must be identified first. Most sheet metal forming models are based on finite element analysis, which are very complex and, therefore, are not suitable for controller design (Majlesi et al. 1992). A piecewise linear model for controller design has been developed in (Majlesi et al. 1992). However, this model cannot be used in closed-loop simulation, because it cannot capture the characteristic nonlinear behavior of a sheet metal forming process. Issues in modeling for control of sheet metal forming have been more fully addressed in (Hsu et al. 2000a, b), especially, from a control point of view. Methods of system identification have been well developed (Ljung 1999) and can be applied to stamping process modeling once a suitable model structure is established (Lim et al. 2010).

The most popular structure for the process controller itself is a proportional-plus-integral controller (Hardt and Fenn 1993; Siegert et al. 1995; Sim and Boyce 1992). However, controller parameters are typically determined by trial and error (Morari and Zafiriou 1989). Although design of process controllers has been well developed (Hsu et al. 1999, 2002; Lim et al. 2010, 2012), its application to sheet metal forming is still being investigated. The reference trajectory in process control is also important to ensure good part quality in sheet metal forming (Hsu et al. 2000b). The reference trajectory has typically been determined experimentally or numerically (Hardt and Fenn 1993; Sim and Boyce 1992), often based upon operator experience. However, optimization of the reference trajectory has not been well addressed (Hsu et al. 2000b, 2002).

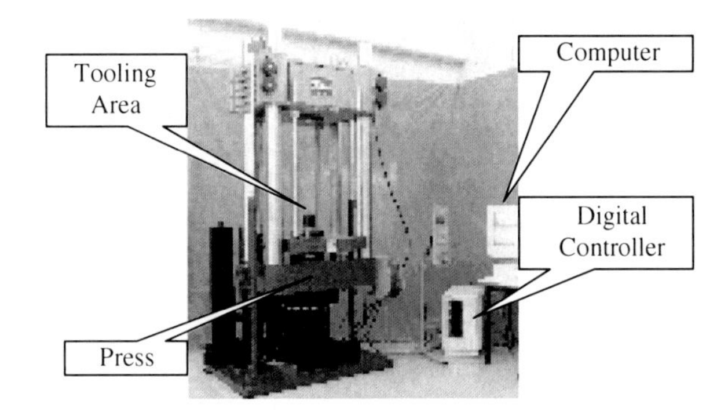

Fig. 5.2 Experimental forming simulator

Key issues regarding the application of process control to sheet metal forming include process modeling for controller design, design of an appropriate process controller and design of an optimal reference trajectory. The purpose of this chapter is to address these key issues to systematically design and implement process control in sheet metal forming based on laboratory experiments using a forming simulator (Hsu et al. 2002).

5.2 Experimental Facility and Model Development

Process control experiments were conducted on a double action laboratory hydraulic forming simulator equipped with a proportional-integral-derivative (PID) digital controller as shown in Fig. 5.2. The press load capacity is 680 kN for the punch and 700 kN for the binder. The digital controller allows the blank holder force, F_b, to track a predetermined reference trajectory, F_{bc}. Thus, this digital PID controller is the realization of the "Machine Controller" block in Fig. 5.1. Implementation of process control on this forming simulator is achieved in the workstation computer as shown in Fig. 5.3 (Hsu et al. 1999, 2000a, 2002). The component labeled "DAQ" is a data acquisition board. It acquires data (i.e., punch force F_p) from the digital controller (realization of the outer feedback path in Fig. 5.1) and feeds the calculated blank holder force command, F_{bc}, to the digital PID machine controller. The "Program" block together with the "DAQ" block in Fig. 5.3 is the realization of the "Process Controller" block in Fig. 5.1. The "WSCI" block is the original workstation communication interface.

A comparison of machine and process control for U-channel forming has demonstrated the superiority of process control over machine control only (Hsu et al. 1999, 2002). Figure 5.4 shows relative tracking errors for machine and process control under dry and lubricated conditions. The results show that process control can maintain the same punch force trajectories under different lubrication conditions while machine control cannot. Table 5.1 shows average measured channel heights for the cases shown in Fig. 5.4. The measurements show that

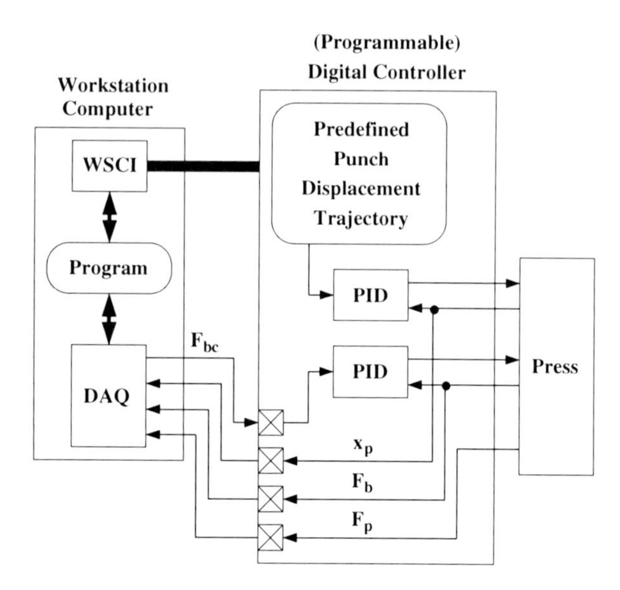

Fig. 5.3 Experimental implementation of the process controller

process control improves consistency in channel height, despite change in lubrication. Therefore, consistency in channel height can be related to consistency in punch force trajectories.

5.3 Establishing the Reference Punch Force

The importance of the reference punch force can be shown by comparing measured channel heights for different reference punch force trajectories (Hsu et al. 2000b, 2002). Figure 5.5 shows two experimental reference punch force trajectories. Table 5.2 shows measured channel heights for these two trajectories. Trajectory (b) produces better parts because springback is minimized and the measured channel heights are closer to the desired channel height (50 mm).

5.4 Process Controller Design

Based on the above experimental results, two important considerations emerge:

- Evaluation of the tracking performance of the process controller.
- Selection of the reference punch force trajectory.

These two considerations will be addressed here. Modeling a sheet metal forming process involving hydraulically controlled single cylinder binder for

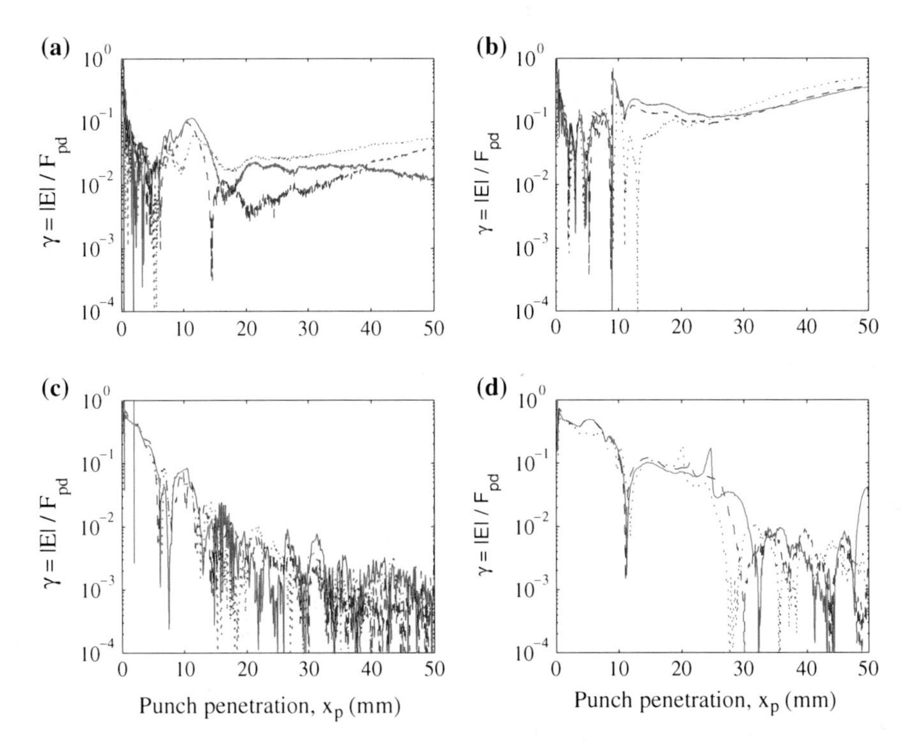

Fig. 5.4 Relative tracking errors. **a** Machine control/dry. **b** Machine control/MP−404. **c** Process control/dry. **d** Process control/MP−404

Table 5.1 Average measured channel heights (mm) for machine and process control under different lubrication conditions

Control type\lubrication	Dry	MP-404
Machine	47.600	46.211
Process	47.557	47.659

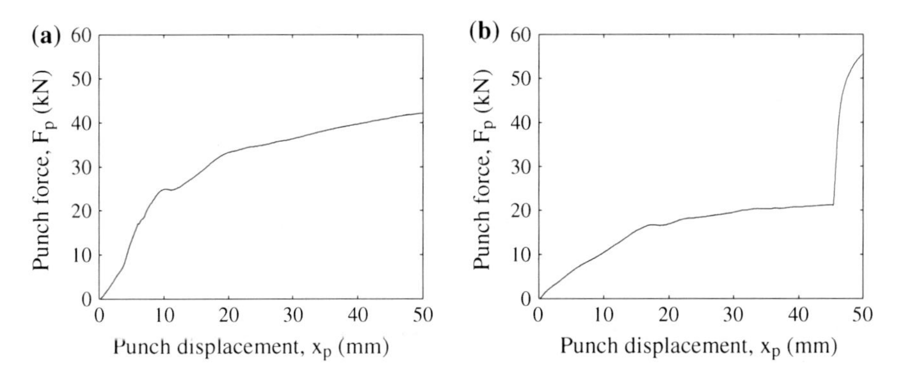

Fig. 5.5 Experimental reference punch force trajectories

Table 5.2 Measured channel heights (mm) for reference punch force trajectories in Fig. 5.5

Trajectory	(a)	(b)
Test #1	47.447	49.251
Test #2	47.396	49.327
Test #3	47.828	49.276
Mean	47.557	49.285

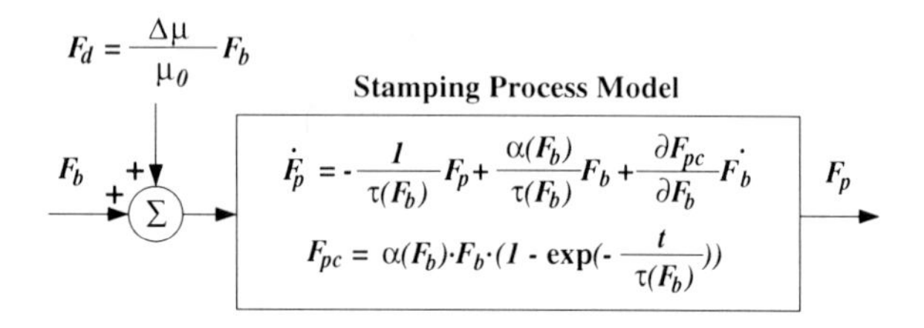

Fig. 5.6 Process model of sheet metal forming

process controller design, which is a single-input single-output (SISO) system, has been investigated (Hsu et al. 2000a). The results of that study are shown in the block diagram in Fig. 5.6. The process model is a first-order nonlinear dynamic model. The disturbance, mainly due to variations in lubrication, is also shown. While this first order dynamic model is nonlinear, it can be linearized about a nominal constant value of the blank holder force, F_{b0} to obtain the response F_{pc}, which leads to a simple control design model as shown in the block diagram in Fig. 5.8. The gain, $\alpha(F_b)$, and time constant, $\tau(F_b)$, of this control-design model depend on the input, F_b. However, if changes in the blank holder force, F_b, are relatively small about the nominal value, F_{b0}, this simple model will be adequate. If not, then an adaptive process controller will be needed to handle the varying gain and time constant in the linearized model.

This model has been successfully used for the U-channel forming process for this laboratory forming simulator (Hsu et al. 2000a). Figure 5.7 shows a comparison of simulation and experimental results for different continuously variable blank holder force trajectories.

Because of the empirically derived process model, systematic study of process controller design can be conducted analytically and numerically before implementation (Hsu et al. 1999, 2002). For this SISO system, a proportional plus integral controller with feedforward action (PIF) has been investigated and successfully implemented in the forming simulator (Hsu et al. 1999, 2002). The block diagram of the controller is shown in Fig. 5.8. A first-order linear model with constant coefficients (i.e., gain and time constant) can be used to design the controller gains. The first-order linear model can then be replaced with the first-order nonlinear model in Fig. 5.6 to evaluate the tracking performance of the closed-loop system using the designed controller gains.

Figure 5.9 shows simulation results with the PIF process controller and the first-order nonlinear model. Figure 5.9a shows the blank holder force automatically generated by the PIF process controller. Figure 5.9b shows the reference punch

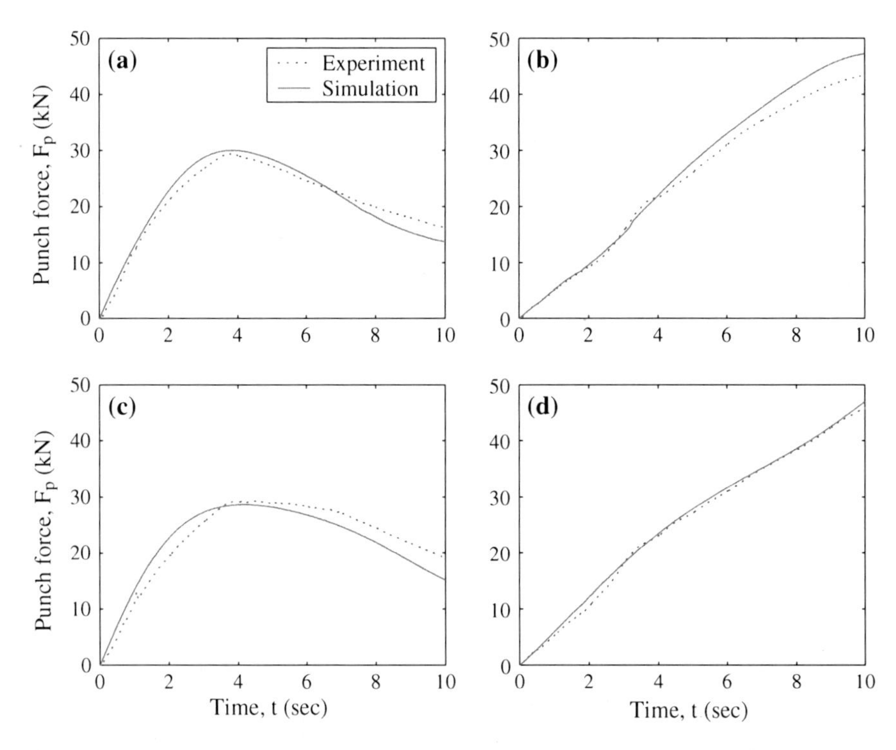

Fig. 5.7 Experimental and predicted punch force trajectories for different variable blankholder force trajectories

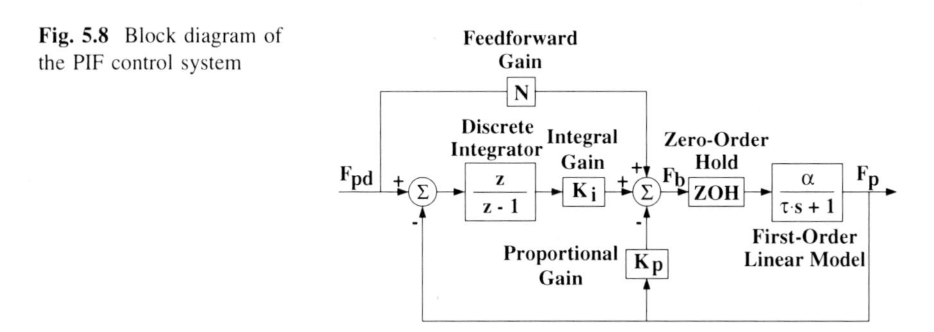

Fig. 5.8 Block diagram of the PIF control system

force trajectory, F_{pd}, and the punch force trajectory, F_p. Good tracking performance can be expected based on these simulation results.

Experimental results using the same PIF process controller and the same reference punch force trajectory are shown in Fig. 5.10. Although there was variation in the blank holder force trajectories, the punch force trajectories were similar. This indicates that the process controller works well despite unmodeled disturbances such as lubrication differences.

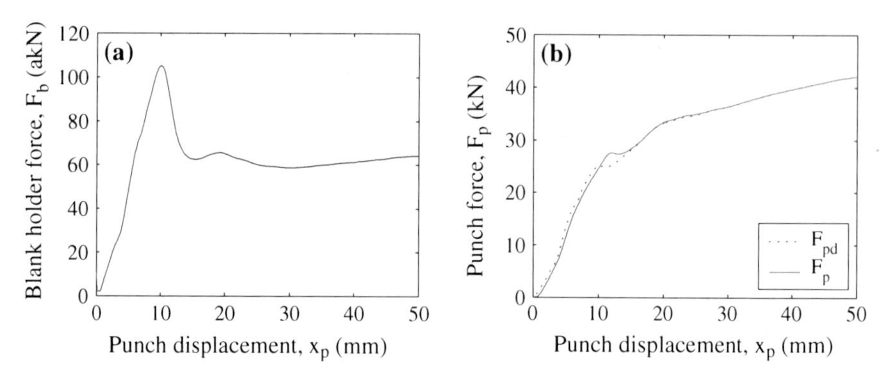

Fig. 5.9 Simulation results using the PIF process controller and the first-order nonlinear model

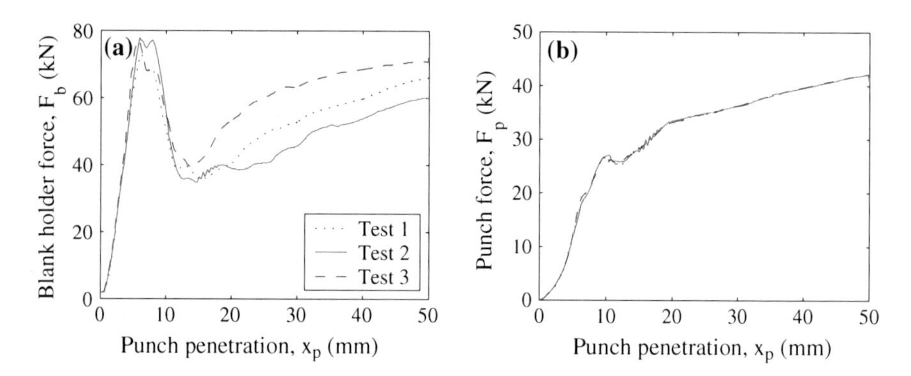

Fig. 5.10 Experimental results using the same PIF process controller and reference punch force trajectory

5.5 Punch Force Reference Trajectory Design

One method for obtaining an optimal reference punch force trajectory is to use design optimization methods (Hsu et al. 2000b; Montgomery 1997). With an ideal process controller, Fig. 5.1 can be simplified as shown in Fig. 5.11.

In this case, the stamped part shape, S, will be totally determined by the reference punch force trajectory, or equivalently, by the punch force trajectory, F_p. A mathematical expression can be used to describe the relationship in Fig. 5.11:

$$S = P(F_p) \tag{5.1}$$

The optimal punch force trajectory F_p^* for a desired shape S_d can be obtained by solving Eq. (5.2):

$$F_p^* = \arg \min_{F_p \in D} E(P(F_p), S_d) \tag{5.2}$$

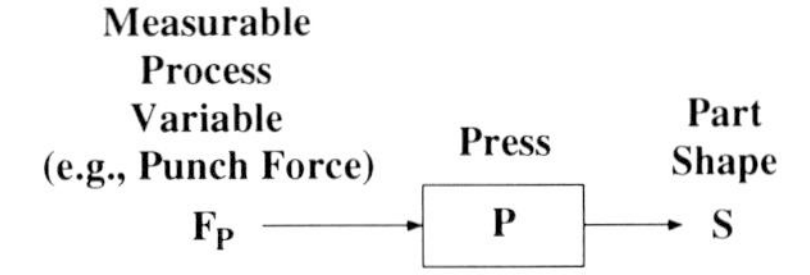

Fig. 5.11 Press with ideal process controller

where F_p^* is the optimal punch force trajectory, D is the safe domain for F_p without tearing and wrinkling, and E is the error cost function used to represent the difference between $P(F_p)$ and S_d.

To find F_p^* through optimization is still difficult. The challenges are:

1. To find the operator P which, given a punch force trajectory, yields the part shape.
2. To find the domain D which defines safe punch force trajectories.

Since current mathematical modeling of sheet metal forming uses finite element methods (Wang and Budinsky 1978; Wenner 1997), there is no simple expression for P or D. A procedure for solving Eq. (5.2) through parameterization and design of experiments can be used as follows:

1. Parameterize F_p and S. Parameters of F_p are the design variables and parameters of S are the response variables.
2. Identify an empirical relationship between the design and response variables.
3. Find the optimal design variables based on the empirical relationship. The optimal punch force trajectory corresponds to the optimal design variables.

Central composite design can be used for design of experiments to fit a second-order model (Montgomery 1997). Response surface methodology can also be used to find the optimal design variables. The methodology is summarized below, and a more detailed description can be found in (Hsu et al. 2000b, 2002).

Typically the smoother the optimal punch force trajectory is, the easier the process controller design is. Parameterization of F_p and S is realized by series expansion with orthogonal functions (e.g., Chebyshev polynomials). The desired smoothness of the optimal punch force trajectory can be ensured by the smoothness of the orthogonal functions.

The above procedure is a sequential one. The following results are from the application of the procedure to U-channel forming. The response variable is the channel height error, e_h, which is defined as the desired channel height minus the measured one. The punch force is parameterized through

$$F_p = a_1\varphi_1 - 2.04\varphi_3 + 5.03\varphi_5 - 1.69\varphi_7 \tag{5.3}$$

where a_1 is the design variable and φ_i is the ith order Chebyshev polynomial. Coded design variables are usually used in design of experiments. The coded design variable, x_1, is

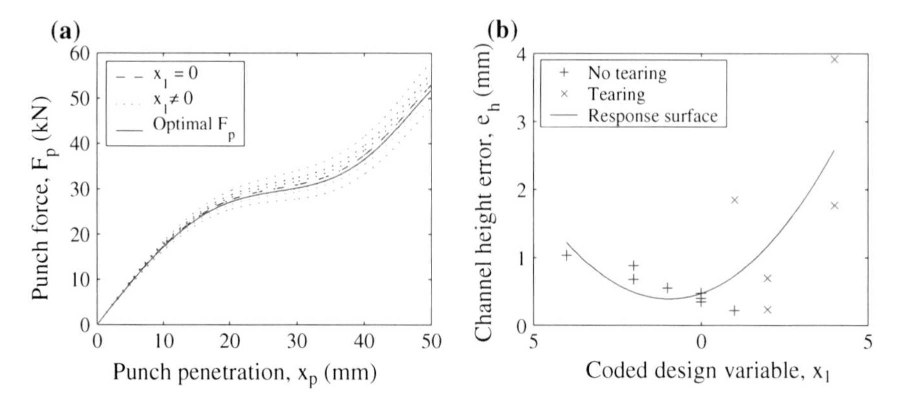

Fig. 5.12 **a** Designed punch force trajectories for the experiments and the optimal F_p. **b** Measured channel height errors and the fitted response surface

$$x_1 = \frac{a_1 - a_{10}}{\lambda a_{10}} \tag{5.4}$$

where a_{10} is the center of the design domain and λ is a scale factor. In this case, for example, a_{10} =51.69 and $\lambda = 0.025$.

Designed punch force trajectories corresponding to $x_1 = 4, 2, 1, 0, -1, -2$, and -4 for the experiments are shown in Fig. 5.12a. Channel height errors are shown in Fig. 5.12b. When tearing occurs, the channel height is assumed to be the height at failure. The optimal F_p in Fig. 5.12a corresponds to the minimum ($x_1^* = -0.94$) of the fitted response surface in Fig. 5.12b.

From a physical point of view, the true optimum in this case is a boundary optimum. Hence, the fitted response surface cannot predict the true optimum precisely. However, the fact that it is a statistically valid model and has a minimum indicates the existence of a true minimum nearby. Based on engineering judgment for the safety and robustness of the forming process, the optimal punch force trajectory is determined as the one corresponding to $x_1 = 0$.

While the laboratory tests described here to determine optimal punch force trajectories can be useful, in practice experienced press operators can quickly determine near-optimal punch force trajectories by trial and error during the die try out process. This is discussed further in subsequent chapters.

5.6 Concluding Remarks

Process control has been shown to improve part quality and consistency in the presence of process disturbances such as varying lubrication conditions. Key issues such as process controller and optimal punch force trajectory design have been addressed. A systematic approach to the application of process control to U-channel forming using a laboratory forming simulator has been presented. A process controller with good tracking performance and an optimal punch force trajectory have been developed.

While these results are important for demonstrating the key concepts of stamping process control, they also show that for practical implementation of these concepts in industrial stamping presses further work is needed. First, the SISO process controller demonstrated here on a laboratory forming simulator must be extended to multi-input multi-output (MIMO) process control for complex parts and industrial presses. Such an extension requires the design of a system with multiple measurements (e.g., punch force at multiple locations on the press) as well as the control of the blank holder force at multiple locations around the die. Second, convenient approaches to process modeling, and to determining the reference punch force trajectories, must be developed. Ideally, such approaches will require a very few (e.g., one) experiments to obtain the data needed for modeling, then automatically generate the required model. Third, adaptive process controllers will be needed to handle the varying parameters of the linearized controller design models for the process. Such adaptive controllers can improve performance, but must be carefully designed to avoid stability problems. Additional issues include the need for fast and accurate machine control (i.e., the inner loop in Fig. 5.1), methods for quickly and easily tuning the process controller gains, etc. The remaining chapters of the book address these and related issues and provide results from die try out and production tests.

References

Adamson, A., Ulsoy, A. G., Demeri, M. Y. (1996) Dimensional control in sheet metal forming via active binder force adjustment. SME Transactions 24:167–178

Ahmetoglu, M., Broek, T. R., Kinzel, G., and Altan, T. (1995) Control of blank holder force to eliminate wrinkling and fracture in deep-drawing rectangular parts. CIRP Annals 44(1):247–250

Hardt, D. E. and Fenn, R. C. (1993) Real-time control of sheet stability during forming. ASME Journal of Engineering for Industry 115:299–308

Hsu, C.-W., Demeri, M. Y., and Ulsoy, A. G. (1999) Improvement of consistency in stamped part quality using process control, sheet metal forming technology. Demeri, M. Y. (ed), TMS Annual Meeting, 53–76.

Hsu, C.-W., Ulsoy, A. G., Demeri, M. Y. (2000a) An approach for modeling sheet metal forming for process controller design. ASME J. of Manufacturing Science and Engineering 122(4): 717–724

Hsu, C.-W., Ulsoy, A. G., and Demeri, M. Y. (2000b) Optimization of the reference punch force trajectory for process control in sheet metal forming. Proceedings of the 2000 Japan-USA Symposium on Flexible Automation Conference, 2000 JUSFA-13139.

Hsu, C.-W., Ulsoy, A. G., Demeri, M. Y. (2002) Development of process control in sheet metal forming. J. of Materials Processing Technology 127(3):361–368

Lim, Y., Venugopal, R., Ulsoy, A. G. (2010) Multi-input multi-output (MIMO) modeling and control for stamping. ASME J. Dynamic Systems, Measurement and Control 132(4):041004 (12 pages)

Lim, Y., Venugopal, R., Ulsoy, A. G. (2012) Auto-tuning and adaptive stamping process control. Control Engineering Practice 20(2):156–164

Ljung, L. (1999) System identification: theory for the user. Prentice-Hall.

Majlessi, S. A., Kashani, A. R., and Weinmann, K. J. (1992) Stamping process model for real-time control. Transactions of NAMRI/SME 1:33–38

Montgomery, D. C. (1997) Design and analysis of experiments, John Wiley & Sons.

Morari, M. and Zafiriou, E. (1989) Robust process control, Prentice-Hall.

Schmoeckel, D. and Beth, M. (1993) Springback reduction in draw-bending process of sheet metals. CIRP Annals 42(1):339–342

Siegert, K., Dannenmann, E., Wagner, S., and Galaiko, A. (1995) Closed-loop control system for blank holder forces in deep drawing. CIRP Annals 44(1):251–254

Siegert, K., Hohnhaus, J., and Wagner, S. (1998) Combination of hydraulic multipoint cushion system and segment-elastic blankholders. SAE 980077

Siegert, K., Ziegler, M., and Wagner, S. (1997) Loop control of the friction force: deep drawing process. Journal of Materials Processing Technology 71:126–133

Sim, H. B. and Boyce, M. C. (1992) Finite element analysis of real-time stability control in sheet metal forming processes. ASME Journal of Engineering Materials and Technology 114:180–188

Wang, N.-M. and Budiansky, B. (1978) Analysis of sheet metal stamping by a finite-element method. ASME Journal of Applied Mechanics 45:73–82

Wenner, M. L. (1997) State-of-the-art of mathematical modeling of sheet metal forming of automotive body panels SAE 970431

Chapter 6
Process Control

Abstract The binder force in sheet metal forming controls the material flow into the die cavity. Maintaining precise material flow characteristics is crucial for producing a high-quality stamped part. Process control can be used to adjust the binder force based on tracking of a reference punch force trajectory to improve part quality and consistency. The purpose of this chapter is to present a systematic approach to the design and implementation of a suitable MIMO process controller. An appropriate process model structure for the purpose of controller design for the sheet metal forming process is presented and parameter estimation for this model is accomplished using system identification methods. This paper is based upon original experiments performed with a new variable blank holder force (or variable binder force) system that includes 12 hydraulic actuators to control the binder force. Experimental results from a complex-geometry part show that the MIMO process controller, designed through simulation is effective.

6.1 Process Modeling and Control

Die design, using the finite element method (FEM) and die try-out, which involves grinding and welding of the die to ensure that the parts produced meet specifications, are time-consuming tasks. Moreover, engineers in the forming industry also face challenging production problems due to process variability. To improve part quality (e.g., eliminating wrinkling, tearing, and springback), with given materials and a conventional press, the original die dimensions based on the part geometry data (e.g., product shape designed by a product designer using computer-aided-design tools) are changed (e.g., working the die/binder geometry or draw-bead) (Sklad et al. 1992). Both die design and die try-out depend heavily on the experience of experts (Manabe et al. 2002).

Controlling the flow of sheet metal via controllable multi-cylinder blank holder actuators reduces die-try out time by cutting down on die rework (e.g., grinding and welding) (Kergen and Jodogne 1992; Siegert et al. 1997, 1998; Lo et al. 1999;

Y. Lim et al., *Process Control for Sheet-Metal Stamping*,
Advances in Industrial Control, DOI: 10.1007/978-1-4471-6284-1_6,

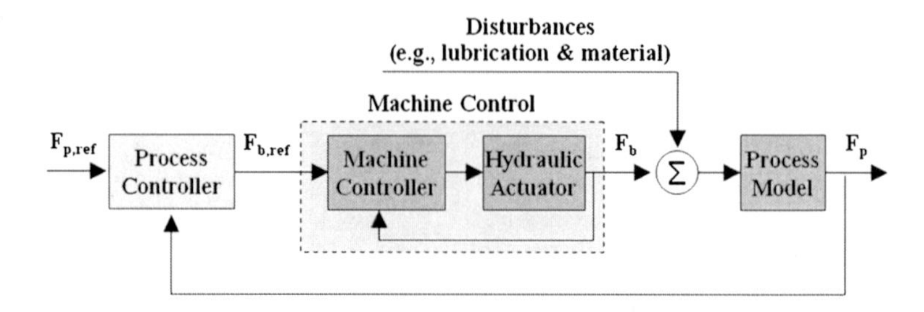

Fig. 6.1 Process control of sheet metal forming

Hsu et al. 2002; Doege et al. 2001, 2002, 2003). Researchers have developed different types of active blank holder systems (e.g., segmented/pulsating blank holder system and reconfigurable discrete die) to improve stamped part quality in forming (Michler et al. 1994; Ziegler 1999; Doege and Elend 2001).

A press with a computer-controlled hydraulic blank holder is capable of controlling the binder force to track a predetermined blank holder force trajectory during forming. As shown by the inner loop in Fig. 6.1, this type of control is referred to as "open-loop" or "machine" control as discussed in Chap. 4. Previous research has shown that machine control can improve material formability, reduce springback, and improve part consistency (Adamson et al. 1996, Sunseri et al. 1996) and can be combined with FEM approach to determine desirable blank holder force references (Sheng et al. 2004; Sunseri et al. 1996; Wang and Lee 2005). However, machine control cannot maintain performance with regard to disturbances occurring during production. Such disturbances can include change in material properties (e.g., formability, blank size, and sheet thickness), change in tooling (e.g., die wear), and variation in lubrication (Hardt 1993; Hardt and Fenn 1993; Hsu et al. 1999; Yagami et al. 2004).

As illustrated in Fig. 6.1, a measurable process variable (e.g., punch force, F_p) is made to track a reference trajectory (i.e., reference punch force, $F_{p,ref}$) through manipulation of a control variable (i.e., binder force, F_b). The process controller is designed to automatically generate the necessary binder force command (i.e., $F_{b,ref}$) for the machine controller to maintain the tracking error between F_p and $F_{p,ref}$ as small as possible in the presence of disturbances. Thus, the closed-loop system, including the process model and the process controller, can achieve high performance tracking of the reference punch force trajectory through manipulation of the binder force regardless of the disturbances. Previous work has shown the effectiveness of process control in sheet metal forming. For example, single-input single-output (SISO) process modeling and control using a proportional plus integral (PI) controller was investigated based on simple die geometry (e.g., *u*-channel forming) under laboratory-based tests (Siegert et al. 1997, 1998; Bohn et al. 1998, Ziegler 1999; Hsu et al. 2000, 2002). However, multi-input multi-output (MIMO) process modeling and control for complex-geometry parts for high-volume production has not previously been studied.

6.2 System Identification

6.2.1 Experimental System

The experimental system, with 12 hydraulic actuators placed underneath the binder, and an Opal-RT real-time data acquisition and control system, is deployed to perform the experiments. When the punch compresses the hydraulic actuator rods located at the bottom of the die (see Fig. 6.2a), blank holder forces acting on the binder of the die at different locations are generated based on feedback pressure measurements using hydraulic pressure sensors located on the hydraulic regulator unit at the back of the die (see Fig. 6.2a). The complex part used for the experiments is a double-door of a pick-up truck made from a tailor-welded steel blank with three different thicknesses (see Fig. 6.2b). The press is a 1,000-ton mechanical press, which can operate at 12 strokes/min. The material flow is controlled by a set of blank holders with the 12 hydraulic actuators. The punch force at the four corners of the press is measured using full-bridge strain gauges, which are attached to the surface of the four punch-supporting beams on the press. The real-time system plays a key role in controlling the system operation and acquiring the measured data from the sensors. The experimental conditions and sensors used in tests are given in Table 6.1.

6.2.2 Experimental Data for System Identification

System identification is an experimental approach to plant modeling. We present a simple mechanics-based approach to establish a potential controller-design model *structure,* for the purpose of controller design. Then, we utilize data obtained from experiments to estimate the unknown parameters in the model. The objective here is to parameterize transfer function models of the forming process, with input–output data from experimental die try-out tests. Thus, it is necessary to integrate in-press sensors and develop real-time data acquisition capabilities to collect data that can be used to parameterize MIMO dynamic models using system identification techniques based on die try-out experiments. In addition, we use such experiments to capture the desired reference trajectories (i.e., punch force trajectories), which characterize a good part.

6.2.2.1 Blank Holder Force Trajectories as Input

Figure 6.3 shows three different types of measured blank holder force trajectories (i.e., $F_b(t)$) for one representative cylinder of the 12 hydraulic actuators. Due to space limitations, only one of the 12 binder force profiles is shown, but the other profiles are qualitatively similar. Trajectory (1) is a *constant* blank holder force

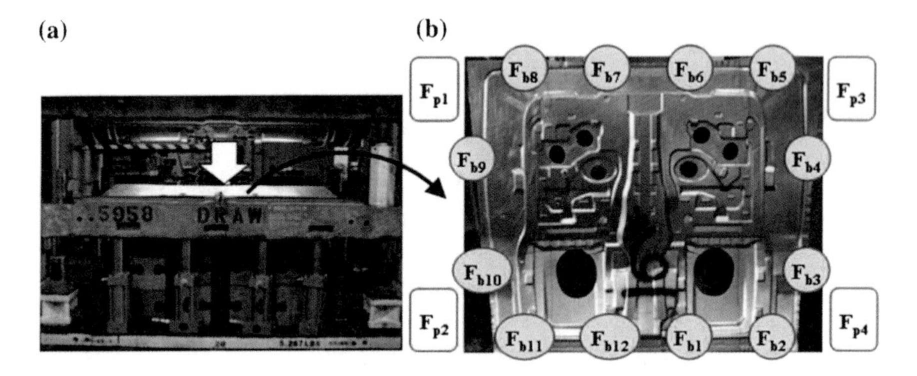

Fig. 6.2 Experimental system: **a** test die with actuators and sensors, **b** a stamped part showing locations of process variables, or *top-view* of (**a**)

Table 6.1 Experimental conditions and sensors for stamping

Punch speed		215 mm/sec
Punch displacement, h_{max}		150 mm
Data sampling rate		0.002 s (500 Hz)
Lubrication		Dry
Material		CR EDD steel
Blank size		1,640 × 1,600 × 0.8 mm
Punch force sensor	Type	Full bridge strain gage
	Excitation	Built-in 10 V@125 mA max
	Accuracy	±1 % of full scale max
Binder force sensor	Type	Piezo-resistive strain gage
	Resolution	1 ~ 5 V for 0 ~ 2,500 psi
Punch stroke sensor	Type	Position transducer
	Accuracy	±0.005 % of full scale

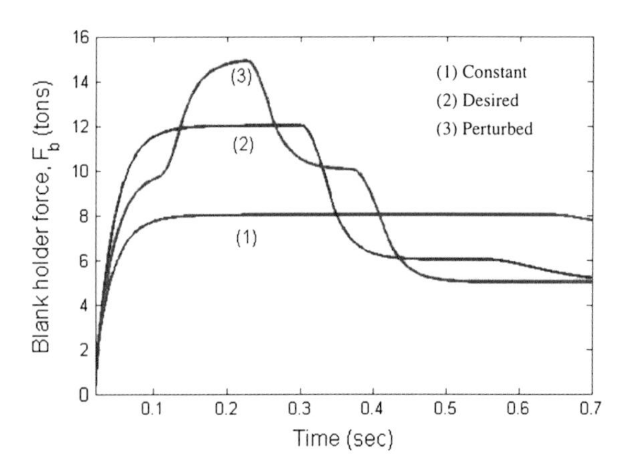

Fig. 6.3 Commanded blank holder force trajectories (i.e., F_{b1}) for a hydraulic actuator shown Fig. 6.2b

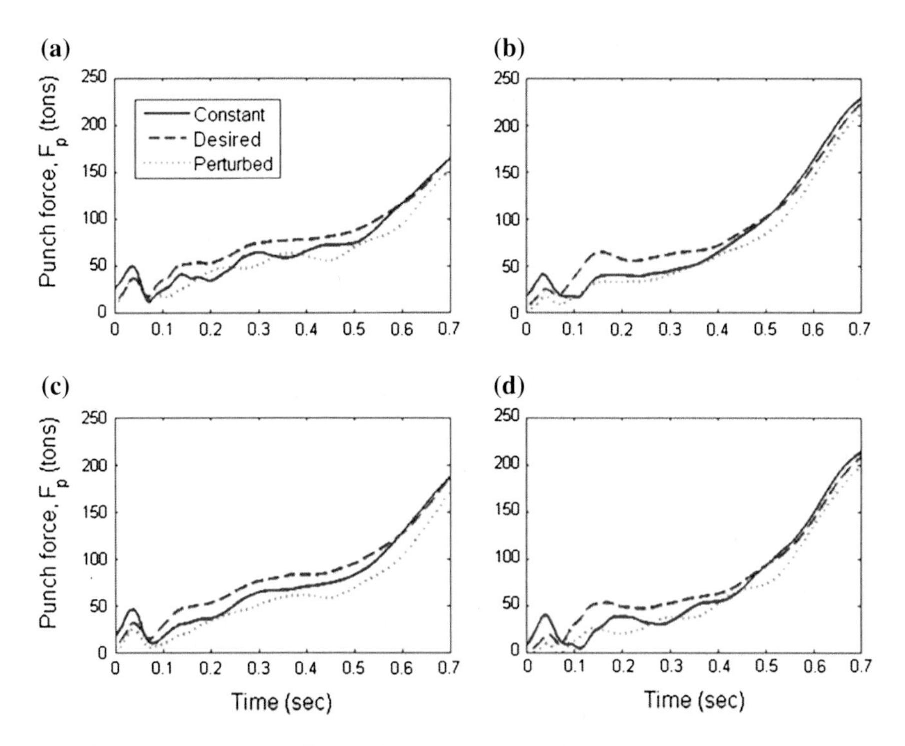

Fig. 6.4 Experimental punch force trajectories based on three different types of blank holder force at four locations: **a** δF_{p1} **b** δF_{p2} **c** δF_{p3} **d** δF_{p4}

experiment used as a baseline. Trajectory (2) is a *desired*, or "optimal", blank holder force trajectory selected by experienced operators for making a good part. Trajectory (3) is a *perturbed* blank holder force (with respect to the baseline) used for system identification.

6.2.2.2 Punch Force Trajectories as Output

Figure 6.4 shows experimental results for four punch force trajectories based on the three different types of blank holder force trajectories. The punch force trajectories obtained, with the *desired* blank holder force, will be used as the reference punch force trajectories in the subsequent design of a process controller. Note, from Fig. 6.4, that the punch force generally increases as the punch stroke increases. Higher blank holder force (or smaller draw-in) produces higher punch force. For example, as shown Figs. 6.3 and 6.4d, during 0–0.4 s, the punch force (F_{p4}) in the desired case is larger than in the constant case, because the binder force (F_{b1}) in the desired case is larger than in the constant case. Also, during 0.4–0.7 s, the same trend is validated; a smaller blank holder force results in a smaller punch force. This input–output relationship will be further investigated in the following process modeling and parameter identification sections. We note that small fluctuations are observed in Fig. 6.4 at the beginning of the stroke due to the compliance of the hydraulic actuator rods as they contact the binder right after the impact of the punch with the lower binder.

6.2.3 Process Model Structure

For designing the stamping process controller, a simple controller-design model, providing a dynamic relationship between the process input (i.e., binder force) and process output (i.e., punch force) is required. Thus, the goal in modeling is not to develop a model suitable for simulating the stamping process, but rather a simple model structure dynamically relating input and output, with undetermined parameters that can then be experimentally evaluated.

Consider a simple one-dimensional analysis, for a cross section of the sheet metal in tension in a simple stamping process, as the sheet is being pulled into the die by the punch. This is schematically illustrated in Fig. 6.5a, where the binder, with force $F_b(t)$, restricts the flow of material into the die. The punch stroke is denoted by $h(t)$, and pulls the element of material into the die. The contact at the binders inhibits the material flow into the die, due to a friction force $\mu F_b(t)$. Due to the resulting deformation of the material in tension, the material draw-in, $l_s(t)$, is less that $h(t)$, i.e., $(h(t) - l_s(t)) > 0$.

For the plastic deformation of the sheet metal in tension, the stress, σ, and strain, ε, are related by Hollomon (1945)

$$\sigma = K\varepsilon^n\dot{\varepsilon}^m \tag{6.1}$$

where $\dot{\varepsilon}$ is the strain rate, and K, n and m are material constant, work-hardening and strain rate sensitivity respectively. Linearization of Eq. (6.1) about nominal values $(\sigma_0,\ \varepsilon_0, \dot{\varepsilon}_0)$ yields

$$\Delta\sigma \approx K_1\Delta\varepsilon + K_2\Delta\dot{\varepsilon} \tag{6.2}$$

The relationship in Eq. (6.2), together with the element geometry, can be used to determine a restraining force, in terms of the elongation and elongation rate in the element of material, of the form

$$F_r = \alpha\{h(t) - l_s(t)\} + \beta\{\dot{h}((t) - \dot{l}_s(t)\} \tag{6.3}$$

where α and β will depend on K, n, m as well as the element geometry and contact conditions. As shown schematically in Fig. 6.5b, one then obtains a lumped-parameter dynamic model for the draw-in

$$m_s\ddot{l}_s(t) + \beta\dot{l}_s(t) + \alpha l_s(t) = \alpha\dot{h}(t) + \beta h(t) - \mu F_b(t) \tag{6.4}$$

where m_s is an equivalent mass for the element under consideration.

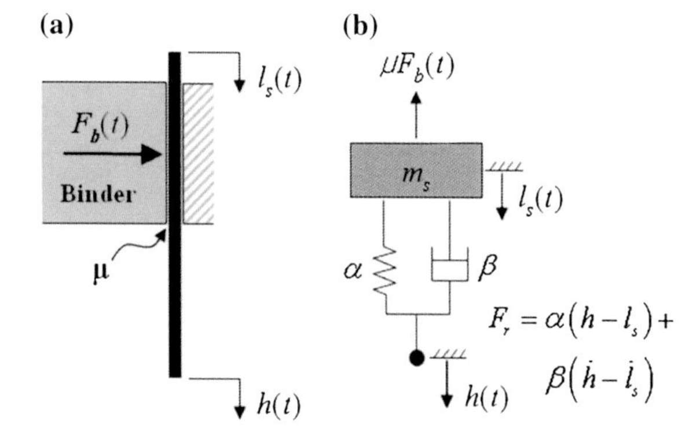

Fig. 6.5 Model for a simple stamping process: **a** schematic **b** lumped model

Consider next the formulation in terms of perturbation variables, which represent changes in these variables from specified values

$$\begin{aligned} h(t) &= h_0(t) + \delta h(t) \\ l_s(t) &= l_{s0}(t) + \delta l_s(t) \\ F_b(t) &= F_{b0}(t) + \delta F_b(t) \\ F_p(t) &= F_{p0}(t) + \delta F_p(t) \end{aligned} \tag{6.5}$$

where $F_p(t)$ denotes the punch force. The punch stroke $h(t)$ is assumed to be prescribed, with no variation, and thus $\delta h(t) = 0$. Also, the restraining force, for the simple one-dimensional geometry being considered, is equal and opposite of the punch force. Thus, $F_p(t) = -F_r(t)$, and by Laplace transformation of Eq. (6.4) one obtains the following transfer function relating the change in binder force (input) to the change in punch force (output)

$$\frac{\delta F_p(s)}{\delta F_b(s)} = \frac{\mu(\beta s + \alpha)}{m_s s^2 + \beta s + \alpha} = \frac{b_1 s + b_0}{s^2 + a_1 s + a_0} \tag{6.6}$$

As illustrated experimentally in Fig. 6.6, a sudden change of input (binder force) and output (punch force) shows qualitative agreement with model structure in Eq. (6.6). A sudden change, in binder force from a nominal value, leads to a change in the punch force with the same sign, a time lag, and overshoot. In the experiment, a small fluctuation in punch force, at the beginning of the stroke (0–0.1 s), occurs due to the impact of the punch with the lower binder, as shown previously in Fig. 6.4.

Thus, Eq. (6.6) provides the basic structure for the required controller-design model, relating the change in the manipulated binder force inputs to the resulting change in the measured punch force outputs. Note, however, that the parameters of the dynamical model in Eq. (6.6) depend on the sheet metal properties and geometry, as well as the die and binder geometry, lubrication, etc. Consequently, the parameters in Eq. (6.6) will need to be experimentally estimated, as described in the next section. Although not presented here, due to space limitations, the basic

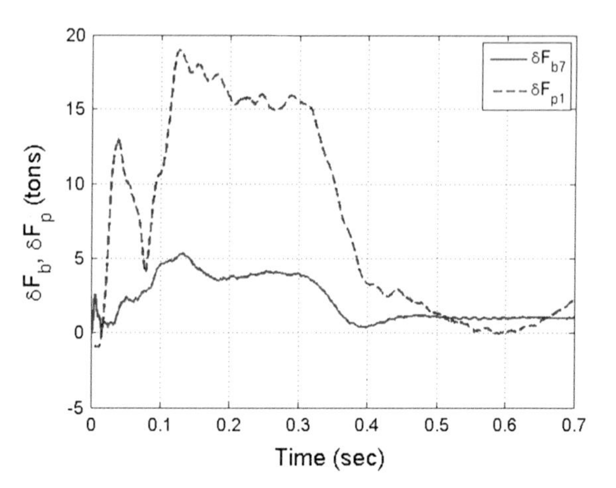

Fig. 6.6 Qualitative validation of the control-design model structure through experiments via comparison to Eq. (6.6) of a change in binder force and the resulting punch force response

relationships between binder force inputs and material draw-in (i.e., Eq. (6.4)) have also been experimentally validated using cable type draw-in sensors.

6.2.4 Parameter Estimation for Process Modeling

Input–output data obtained from die try-out tests are used to parameterize the process models. The MIMO complete process model structure for parameter estimation consists of three separate dynamic models: machine control, process model, and low-pass filter.

Since the process variables are sampled in connection with the analog-to-digital conversion, Eq. (6.6) in continuous-time is converted into a transfer function for a second-order process model (G_p) in discrete-time with sampling rate $Ts = 0.002$ s. The resulting process model structure, which characterizes the simple process dynamics relating the actual binder force ($F_{b,act}$) as an input and the actual punch force ($F_{p,act}$) as an output, is formulated with unknown parameters in discrete-time as

$$G_p(z) = \frac{\delta F_{p,act}(z)}{\delta F_{b,act}(z)} = \frac{b_1 z + b_0}{z^2 + a_1 z + a_0} \tag{6.7}$$

Second, based on experimental observations, the transfer function of the 1st order machine control model (G_m) shown in Fig. 6.7, which characterizes the reference binder force ($F_{b,ref}$) as an input and the actual binder force ($F_{b,act}$) as an output, is formulated as a first order transfer function with unknown parameters in discrete-time as

$$G_m(z) = \frac{\delta F_{b,act}(z)}{\delta F_{b,ref}(z)} = \frac{d_0}{z + c_0} \tag{6.8}$$

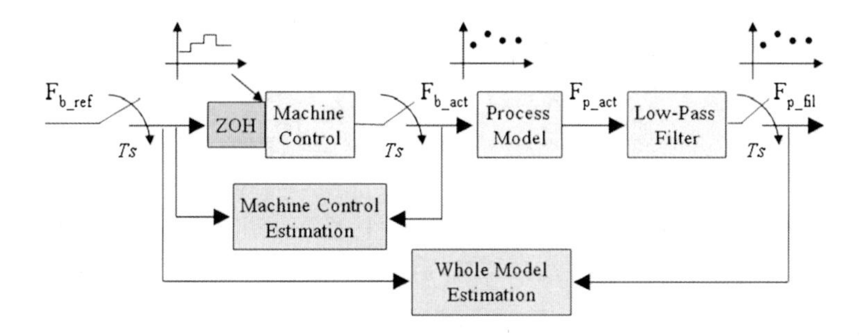

Fig. 6.7 Schematic diagram of process model for estimation ($Ts = 0.002$ s)

The unknown parameters in the machine control models are estimated using two methods: the N4SID subspace algorithm and the standard Least Squares (LS) algorithm (Astrom et al. 1990). The machine control models obtained using these two system identification methods are validated by matching the measured desired binder force output with the actual binder force generated by the identified machine control model with the commanded desired binder force as the input (see Fig. 6.8a–c).

Third, the transfer function of the first order low-pass filter (G_f) shown in Fig. 6.7, is given, with known parameters, in discrete-time as

$$G_f(z) = \frac{\delta F_{p,fil}(z)}{\delta F_{p,act}(z)} = \frac{50z}{z - 0.9048} \tag{6.9}$$

The filter is discretized using an impulse-invariant transformation with a sampling period of 0.002 s, while the machine control model utilizes a zero-order-hold (ZOH) to characterize the digital-to-analog (D–A) converter in the real-time system. As shown in Fig. 6.7, cascading the 3 transfer functions in Eqs. (6.7), (6.8) and (6.9), and using the mean machine control models, the total model structure is formulated in discrete-time with unknown process model parameters as

$$G_{total}(z) = \frac{\delta F_{p,fil}(z)}{\delta F_{b,ref}(z)} = G_p(z) \cdot G_m(z) \cdot G_f(z) \tag{6.10}$$

Moreover, one can extend the model structure in Eq. (6.10) to the MIMO case by creating a 4×12 transfer function matrix (TFM) in Eq. (6.11), with 4-punch force ($\delta F_{p,fil}$) as output and 12-binder force ($\delta F_{b,ref}$) as inputs. Based on the experimentally validated assumption that each punch force output is heavily affected only by the three nearest binder force inputs (see Fig. 6.2b), we constrain the TFM to a block-diagonal form given by

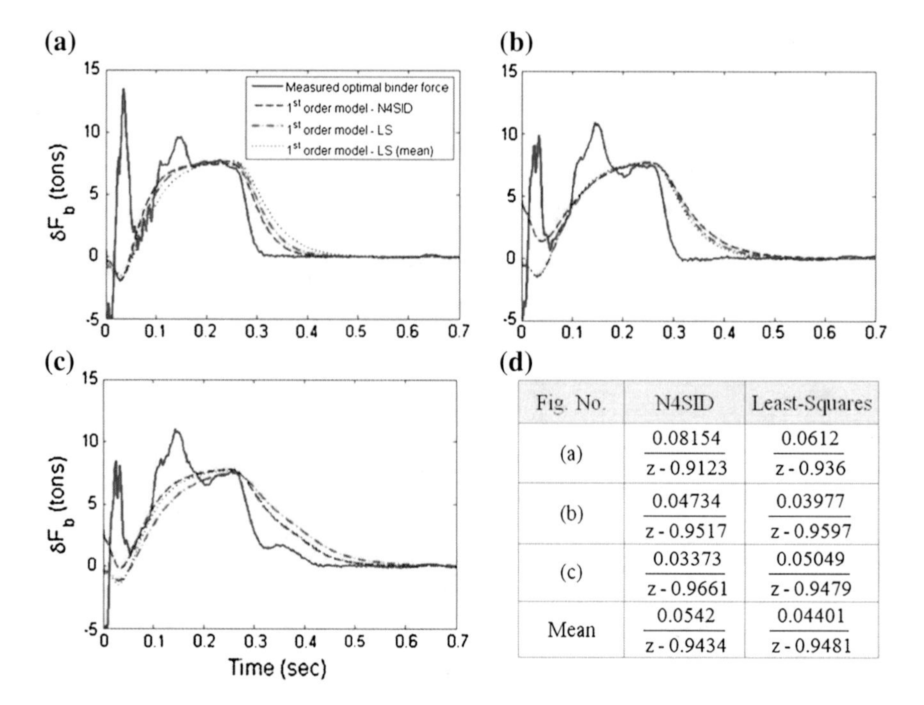

Fig. No.	N4SID	Least-Squares
(a)	$\frac{0.08154}{z-0.9123}$	$\frac{0.0612}{z-0.936}$
(b)	$\frac{0.04734}{z-0.9517}$	$\frac{0.03977}{z-0.9597}$
(c)	$\frac{0.03373}{z-0.9661}$	$\frac{0.05049}{z-0.9479}$
Mean	$\frac{0.0542}{z-0.9434}$	$\frac{0.04401}{z-0.9481}$

Fig. 6.8 Estimated machine control (MC) models obtained by N4SID and least squares algorithm: **a** F_{b7} **b** F_{b8} **c** F_{b9} **d** estimated MC models

$$\begin{bmatrix} \delta F_{p1}(z) \\ \delta F_{p2}(z) \\ \delta F_{p3}(z) \\ \delta F_{p4}(z) \end{bmatrix} = \begin{bmatrix} G_{1,1} & G_{1,2} & G_{1,3} & 0 & 0 & 0 & 0 & 0 & 0 & 0 & 0 & 0 \\ 0 & 0 & 0 & G_{2,4} & G_{2,5} & G_{2,6} & 0 & 0 & 0 & 0 & 0 & 0 \\ 0 & 0 & 0 & 0 & 0 & 0 & G_{3,7} & G_{3,8} & G_{3,9} & 0 & 0 & 0 \\ 0 & 0 & 0 & 0 & 0 & 0 & 0 & 0 & 0 & G_{4,10} & G_{4,11} & G_{4,12} \end{bmatrix} \begin{bmatrix} \delta F_{b7}(z) \\ \delta F_{b8}(z) \\ \vdots \\ \delta F_{b2}(z) \\ \delta F_{b3}(z) \end{bmatrix} \tag{6.11}$$

Experimental data show that this structure is sufficient to characterize the dynamics of the process from actuator reference inputs to filtered sensor outputs. Thus, the structure in Eq. (6.11) represents a collection of four MISO systems (one at each corner) with one punch force output and three binder force inputs. The unknown parameters in the fourth order system models are estimated based on the experimental data using the Least Squares (LS) algorithm and are plotted in Fig. 6.9. Each estimated model characterizes the dynamics of the process from three reference binder force (i.e., $F_{b,ref}$) inputs (as shown in Fig. 6.3) to one filtered punch force (i.e., $F_{p,fil}$) output (as shown in Fig. 6.4). Estimated parameters of the fourth order perturbation process models in discrete-time with respect to the each punch force output are also given in Table 6.2.

As shown in Fig. 6.10 for the model validation, experimental punch forces recorded for the desired case (solid-line in Fig. 6.10) are compared with the punch

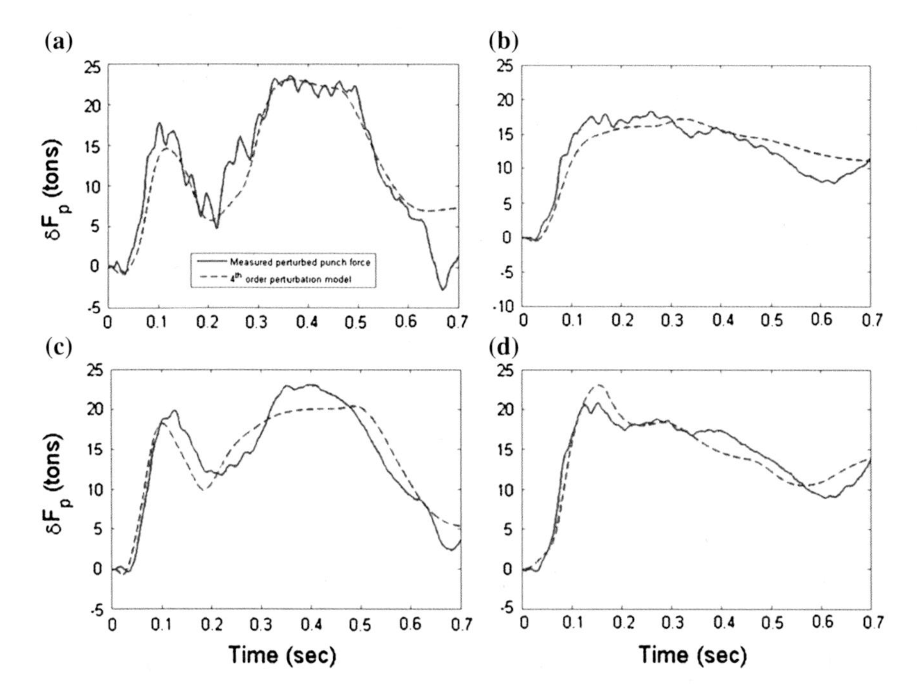

Fig. 6.9 Estimated fourth order perturbation model based on experimental data: **a** δF_{p1} **b** δF_{p2} **c** δF_{p3} **d** δF_{p4}

Table 6.2 Estimated parameters of the fourth order perturbation model in transfer function matrix in Eq. (6.11)

	$\delta F_{p1}(z)$		$\delta F_{p2}(z)$
$G_{1,1}$	$\frac{0.8432z^2-0.83.21\,z}{z^4-0.9746\,z^3+0.3065\,z^2-0.3469\,z+0.1798}$	$G_{2,4}$	$\frac{0.2462\,z^2-0.1468\,z}{z^4-0.7203\,z^3-0.1166\,z^2-0.2076\,z+0.1942}$
$G_{1,2}$	$\frac{2.095\,z^2-1.586\,z}{z^4-0.9746\,z^3+0.3065\,z^2 0.3469z+0.1798}$	$G_{2,5}$	$\frac{0.8538\,z^2-0.7037\,z}{z^4-0.7203\,z^3-0.1166\,z^2-0.2076\,z+0.1942}$
$G_{1,3}$	$\frac{1.526\,z^2-1.479\,z}{z^4-0.9746\,z^3+0.3065\,z^2-0.3469\,z+0.1798}$	$G_{3,6}$	$\frac{0.1353\,z^2-0.0927\,z}{z^4-0.7203\,z^3-0.1166\,z^2-0.2076\,z+0.1942}$
	$\delta F_{p3}(z)$		$\delta F_{p4}(z)$
$G_{3,7}$	$\frac{0.4912\,z^2-0.2313\,z}{z^4-1.403\,z^3+0.3936\,z^2+0.1452z-0.05513}$	$G_{4,10}$	$\frac{0.7219\,z^2-0.5878z}{z^4-1.621\,z^3+0.6777\,z^2+0.06901\,z-0.06487}$
$G_{3,8}$	$\frac{0.333\,z^2-0.03505\,z}{z^4-1.403\,z^3+0.3936\,z^2+0.1452\,z-0.05513}$	$G_{4,11}$	$\frac{0.2685\,z^2-0.1701\,z}{z^4-1.621\,z^3+0.6777\,z^2+0.6901\,z-0.06487}$
$G_{3,9}$	$\frac{0.9474\,z^2-0.66\,z}{z^4-1.403\,z^3+0.3936\,z^2+0.1452z-0.05513}$	$G_{4,12}$	$\frac{0.4632\,z^2-0.4214\,z}{z^4-1.621\,z^3+0.6777\,z^2+0.06901\,z-0.06487}$

force outputs generated by estimated models (dotted-line in Fig. 6.10) using the command binder forces in the desired case. The agreement is acceptable, for our purposes of controller design, but the discrepancies are significant. There are two major reasons why measured punch force outputs are not well matched with outputs from estimated models in these validation results. First, nonlinearity in plastic deformation was ignored in obtaining our process model structure through linearization. Second, although forming is a complex three-dimensional phenomenon, only a simple one-dimensional analysis was considered in developing

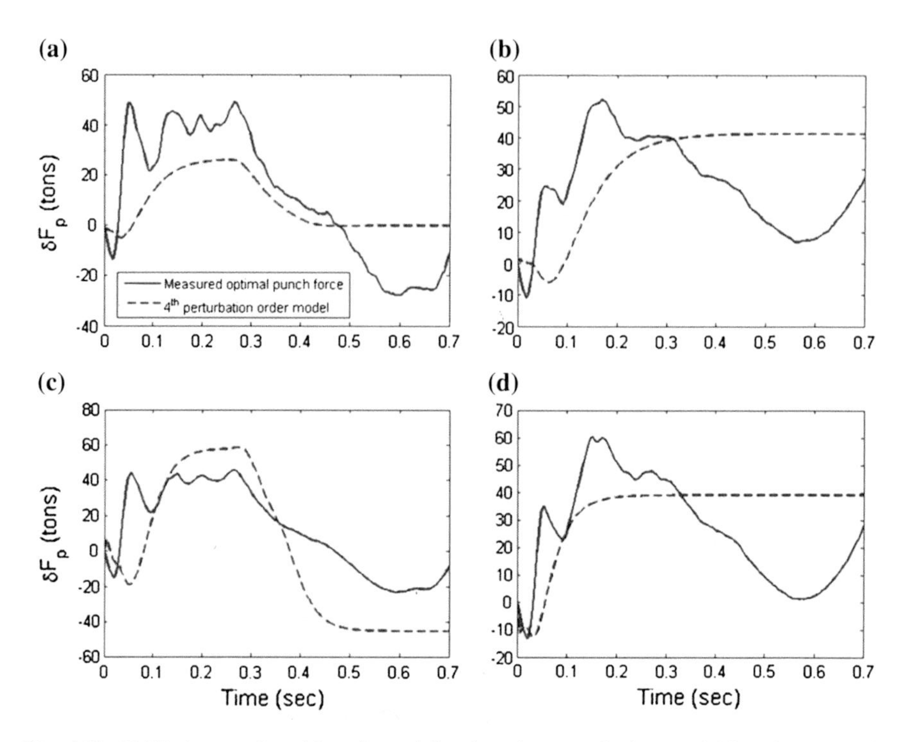

Fig. 6.10 Validation results with estimated fourth order perturbation model based on experimental data: **a** δF_{p1} **b** δF_{p2} **c** δF_{p3} **d** δF_{p4}

the control-design model structure. However, it is shown in the next section that the use of this simple model structure for controller design is adequate for obtaining good closed-loop system performance.

6.3 Process Control Design Based on MIMO Model

The process controller requires high performance tracking of the reference trajectory (e.g., the punch force) through manipulation of the blank holder force in the presence of disturbances, such as lubrication change and blank material property or thickness change (see Fig. 6.1). In other words, the process control is able to maintain the desired punch force trajectories under different lubrication conditions and blank material characteristics while machine control (or an open-loop system) cannot. For example, as illustrated in Fig. 6.11 for a representative test, when there is lubricant on the sheet metal, the punch force output is smaller than under non-lubricated conditions, with the same binder forces conditions. The process controller enables tracking of the reference punch force by adjusting binder forces. Similarly, for the thicker material, the punch force output will

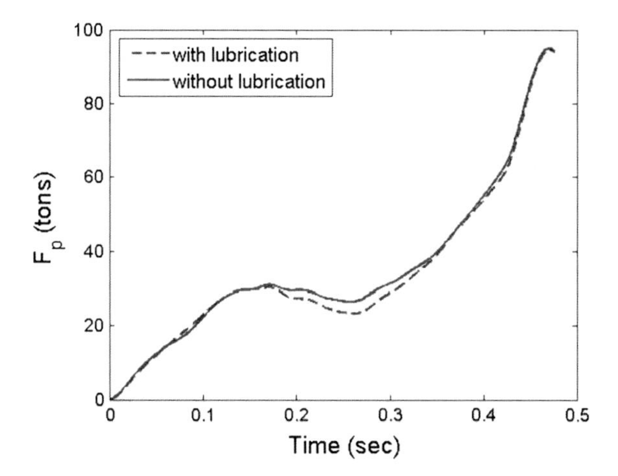

Fig. 6.11 The punch force differences between lubricated and non-lubricated conditions with the same binder forces

increase. However, the process controller enables tracking of the reference punch force and rejects such disturbances by decreasing blank holder forces. The purpose of following section is to present a systematic approach to the design and implementation of a suitable MIMO process controller.

6.3.1 Design of the PI Process Controller

For the MIMO system given by the block-diagonal form in Eq. (6.11), four SIMO proportional plus integral (PI) controllers are implemented using OPAL-RT's RT-LAB and The Mathworks' Simulink/Real-Time Workshop® in the experimental real-time system. The block diagram of the SIMO process controller at each corner is shown in Fig. 6.12. The fourth order estimated perturbation models in Table 6.2 are used to design the PI process controller gains.

To design the SIMO PI controller based on the MISO perturbed process model at each corner, five steps are followed:

- *Step 1*: Determine PI control gains based on a linear process model by using the root-locus design method to evaluate how the PI controller gains influence the closed-loop pole locations.
- *Step 2*: Investigate the gain margin (GM) and phase margin (PM) using a frequency–response design method (e.g., Bode plot). Based on the PI controller gains determined from Step 1, stability margins (i.e., GM and PM) are investigated for several cases described in Table 6.3. It is recommended to provide gain margins not less than 6 dB, and phase margins not less than $\pi/6$ (Safonov 1980). A sample result in Fig. 6.13, shows a controller design where GM is greater than 60 dB and PM is greater than $\pi/2$.

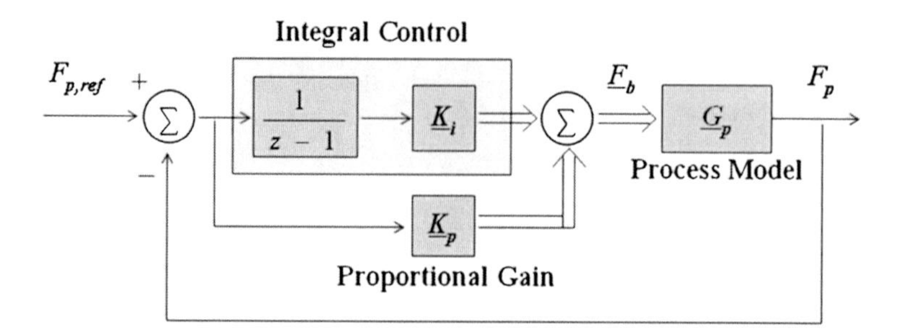

Fig. 6.12 Block diagram of SIMO PI control system at each corner of the press/die

Table 6.3 Cases for simulation and experiment

	Plant model	Cases	$\underline{K}_p$ gains (1×3)	$\underline{K}_i$ gains (1×3)
Simulation	All models $(\delta F_{pj} \sim \delta F_{p4})$	Case I	[0.3 0.3 0.3]	[0.01 0.01 0.01]
		Case II	[0.4 0.4 0.4]	[0.02 0.02 0.02]
		Case III	[0.5 0.5 0.5]	[0.03 0.03 0.03]
Experiment	F_{p1}	Experiment case	[0.4 0.2 0.6]	[0.01 0.01 0.01]
	F_{p2}		[0.1 0.15 0.15]	[0.01 0.01 0.01]
	F_{p3}		[0.6 0.5 0.4]	[0.01 0.01 0.01]
	F_{p4}		[0.15 0.15 0.1]	[0.01 0.01 0.01]

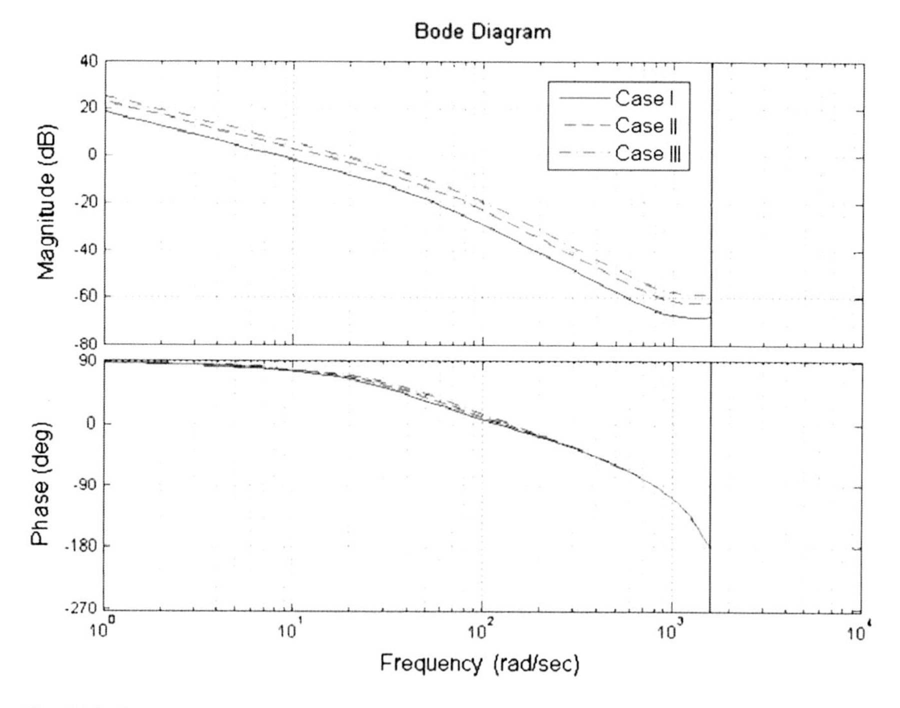

Fig. 6.13 Frequency response analysis based on the PI control gains for the punch force (i.e., F_{p1})

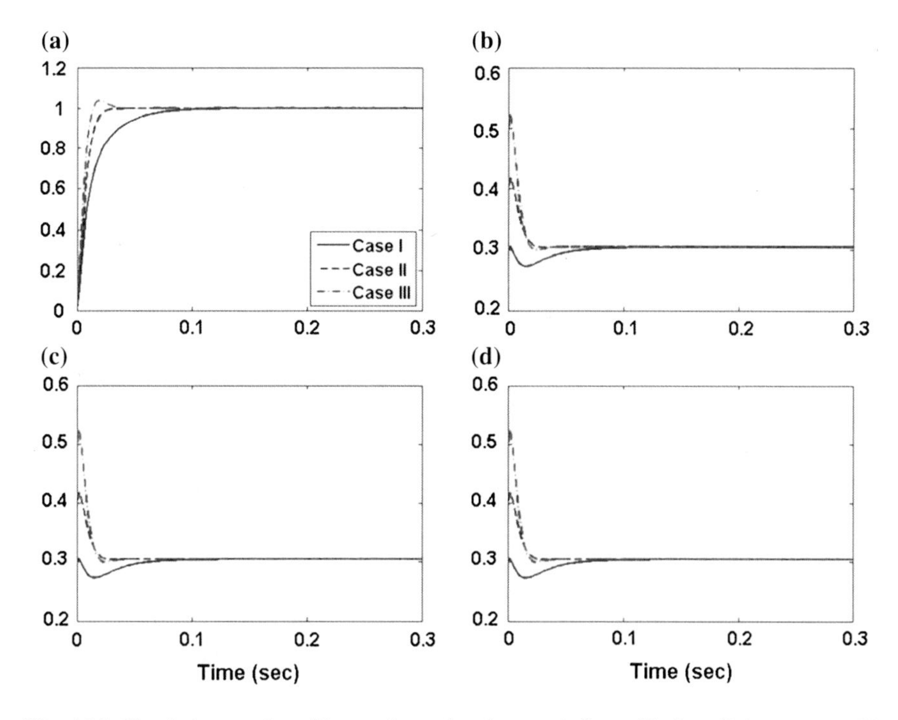

Fig. 6.14 Simulation results with step input for the punch force #1 (i.e., F_{p1}): **a** output #1, **b** input #1, **c** input #2, **d** input #3

- *Step 3*: Check system transient performance (e.g., rise time and settling time) based on three cases of different PI controller gains determined from Step 1, using the closed-loop step response (see Fig. 6.14). For example, settling time of Case I and Case II is less than 0.05 s.
- *Step 4*: Perform simulations based on three cases of PI controller gains, with experimentally determined reference punch forces. This step is used to assess the tracking performance of the controller while ensuring that the control signals meet the binder force saturation constraints (minimum 0 tons and maximum 16 tons).
- *Step 5*: Perform experiments with selected gains for the PI process controller.

6.3.2 Simulation Results with PI Controller Based on Estimated Model

Simulation is used to validate the performance of the proposed PI process controller based on the estimated perturbation process models in Table 6.2. The simulation models use the perturbed binder forces (δF_b) as inputs, the perturbed punch force (δF_p) as output, and the perturbed punch force ($\delta F_{p,ref}$) as the

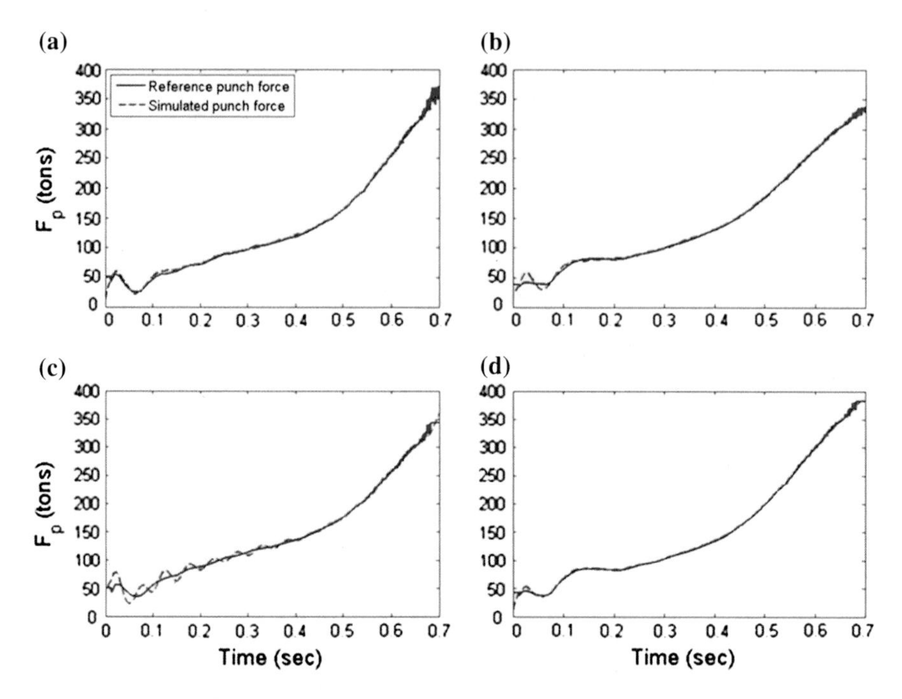

Fig. 6.15 Simulation results of punch force as output tracking reference punch force: **a** F_{p1}, **b** F_{p2}, **c** F_{p3}, **d** F_{p4}

reference. The perturbations are with respect to baseline binder force inputs. The total simulated punch force (F_p), reference punch force ($F_{p,ref}$) and binder forces (F_b), shown in Fig. 6.12, are respectively given by

$$\begin{aligned} F_p &= F_{p,base} + \delta F_p \\ F_{p,ref} &= F_{p,base} + \delta F_{p,ref},\ F_p \& F_{p,ref} \in R^1 \end{aligned} \tag{6.12}$$

$$\underline{F}_b = \underline{F}_{b,base} + \underline{\delta F}_b,\ \underline{F}_b \in R^{1\times 3} \tag{6.13}$$

where $F_{p,base}$ is the measured-baseline of the punch force corresponding to the baseline binder forces, $F_{b,base}$, which is set to a constant value of 16-tons for all 12 actuators (see Fig. 6.2b). Figure 6.15 shows the simulation results for punch force as output using the PI process controller based on the process models. Based on these simulation results, good experimental tracking performance is expected. Simulation and experimental results are compared with respect to the three binder forces associated with each punch force shown in Fig. 6.15. Due to space limitations, simulated and measured binder forces are shown only for the one corner of punch force output (i.e., F_{p1}). Although there are differences between simulated binder forces and experimentally measured binder forces, the simulated binder

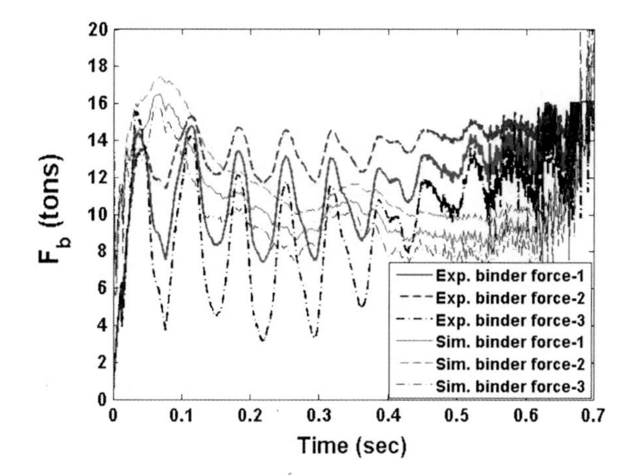

Fig. 6.16 Comparison of three binder forces for the punch force (i.e., F_{p1} shown in Fig. 6.14) between experiment and simulation

force trajectories are similar to the measured binder force trajectories (Fig. 6.16). The high-frequency oscillatory behavior at the end of the punch stroke arises from an implementation issue in reference punch force signal generation which is being corrected for future experiments. However, these oscillations are at a frequency that is sufficiently high as to not affect the stamping process, as evidenced by the fact that they are filtered out by the process dynamics in the punch force output.

6.3.3 Experimental Results

Experimental results using the MIMO PI process controller and the reference punch force trajectories are shown in Fig. 6.17 (see Table 6.1 for experimental conditions and sensors used in tests). As illustrated in Fig. 6.18, it is noted that the process controller (PC) enables accurate punch force output tracking by manipulation of the binder forces. The experimental results shown here demonstrate the effectiveness of the process controller under two extreme conditions. In the first test, all the binder force actuators are initially set to generate 8 tons, a situation in which the total binder force is below the *desired* or *optimal* setting and results in wrinkling, without process control. However, the process controller automatically corrects the binder force based on punch force measurements to eliminate the wrinkles. In the second test, all the binder force actuators are initially commanded to generate 16 tons, a situation in which the total binder force is above *optimal* setting and results in tearing without process control. The rationale behind these tests is that if the process controller is able to correct for these extreme variations,

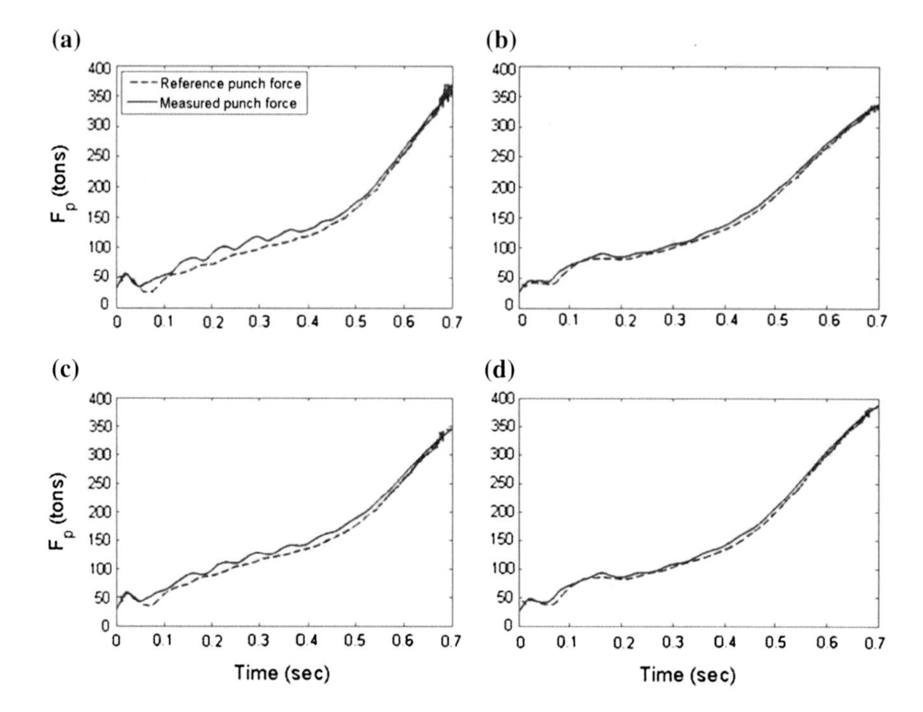

Fig. 6.17 Experimental results of punch force as output tracking reference punch force: **a** F_{p1}, **b** F_{p2}, **c** F_{p3}, **d** F_{p4}

it will be able to correct for the smaller punch-force trajectory deviations observed due to in-process variations arising from disturbances.

The results show that the process control is highly effective in reducing the two problems of wrinkling and tearing. First, Fig. 6.18a illustrates the wrinkling problem; this can occur not only because of low binder forces (i.e., 8 tons), but also because of excessive lubrication allowing too much material flow-in, even with the same blank holder force conditions. Second, Fig. 6.18b shows the tearing problem; this can occur not only because of high binder forces (i.e., 16 tons), but also because of thicker material causing the binder to hold the blank tighter and restrict material flow, as a result of variations in blank sheet production. Thus, although the binder forces have been initially commanded to a constant value of 8 tons (or 16 tons), Fig. 6.18c demonstrates that the MIMO process control adjusts the binder forces to track the reference punch force trajectories, to eliminate wrinkling or tearing. Consequently, the MIMO process control is shown to correct these defects by appropriately regulating the material draw-in in the presence of stamping process disturbances.

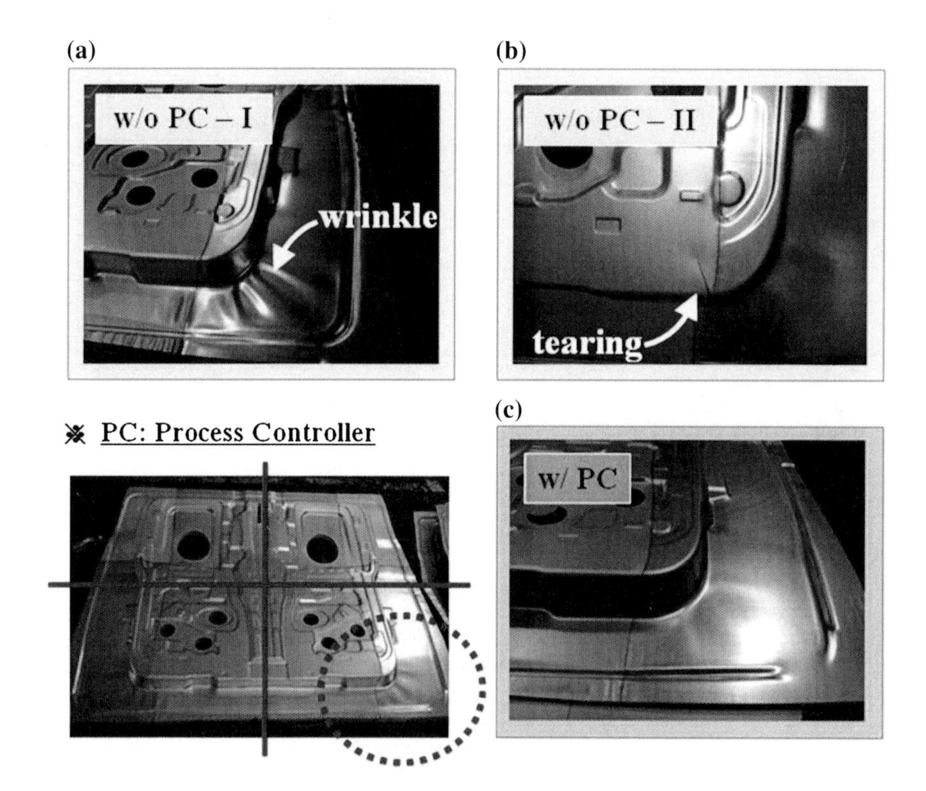

Fig. 6.18 Improved part quality comparisons: **a** wrinkling problem with constant 8 ton binder force, without PC **b** tearing problem with constant 16-ton binder force, without PC **c** improved part, with PC for complex part geometry (i.e., double-door panel)

6.4 Concluding Remarks

In this chapter we describe the development of a simple dynamic input–output controller design model for MIMO process control in stamping. The model structure then provides a basis for estimation of model parameters using system identification techniques. Furthermore, a MIMO stamping process control has been designed and shown to improve part quality and consistency for a complex-geometry part in the presence of plant disturbances. For the first time, a MIMO PI stamping process controller with good tracking performance has been developed and experimentally validated. However, controller fine-tuning based on trial-and-error in experimental tests can be time consuming and expensive. Thus, next chapter presents auto-tuning method for PI process controller. We also describe an adaptive process controller whose parameters are continuously adjusted to accommodate changes in process dynamics and disturbances. The adaptive controller updates the gains of the MIMO PI process controller (which are initially set to the values obtained from auto-tuning) to minimize tracking error, in the presence of plant variations.

References

Adamson, A., Ulsoy, A.G., Demeri, M., (1996). Dimensional Control in Sheet Metal Forming via Active Binder Force Adjustment. SME Transactions. Vol. 24, pp. 167–178.

Astrom, Karl J., Wittenmark, Bjorn, (1990). Computer-Controlled Systems, Prentice Hall, pp. 416–436.

Bohn, Stefen U. Jurthe, Weinmann, Klaus J., (1998). A New Multi-point Active Drawbead Forming Die: Model Development for Process Optimization. SAE Paper No. 980076, pp. 24–30.

Doege, E., Elend, L.E., (2001). Design and application of pliable blank holder systems for the optimization of process conditions in sheet metal forming. J of Materials Proc. Tech., Vol 111, pp. 182–187.

Doege, E., Schmidt-Jurgensen, R., Huinink, S., Yun, J.-W, (2003). Development of an optical sensor for the measurement of the material flow in deep drawing processes. CIRP Annals—Manufacturing Technology, Vol 52, n1, pp. 225–228.

Doege, E., Seidel, H.J., Griesbach, B., Yun, J.W., (2002). Contactless on-line measurement of material flow for closed loop control of deep drawing. J of Materials Proc. Tech., 130–131, pp. 95–99.

Hardt, D. E., (1993). Modeling and Control of Manufacturing Processes: Getting More Involved. Journal of Dynamic Systems, Measurement, and Control, Vol. 115, pp. 291–300.

Hardt, D. E., Fenn, R. C., (1993). Real-Time Control of Sheet Stability during Forming, Journal of Dynamic Systems, Measurement, and Control, Vol. 115, pp. 301–308.

Hollomon, J.H., (1945). Plastic Flow of Metals. Metal Tech., Vol. 12.

Hsu, C.W., Ulsoy, A.G., Demeri, M.Y., (2002). Development of process control in sheet metal forming. J. of Materials Proc. Tech., Vol. 127, pp. 361–368.

Hsu, C.W., Ulsoy, A.G., Demeri, M.Y., (2000). An Approach for Modeling Sheet Metal Forming for Process Controller Design. ASME J. Manuf. Sci. Eng., 122, pp. 717–724.

Hsu, C.W., Ulsoy, A.G., Demeri, M.Y., (1999). Process Controller Design for Sheet Metal Forming. American Control Conference, Vol. 1, pp. 192–196.

Kergen, R., Jodogne, P., (1992). Computerized Control of the Blankholder Pressure on Deep Drawing Process. SAE Paper No. 920433, pp. 51–55.

Lo, Sy-Wei, Jeng, Guo-Ming, (1999). Monitoring the Displacement of a Blank in a Deep Drawing Process by Using a New Embedded-Type Sensor. Int J Adv Manuf Tech., vol. 15, pp. 815–821.

Manabe, K., Koyama, H., Katoh, K., Yoshihara, S., Yagami, T., (2002). Development of a combination punch speed and blank-holder fuzzy control system for the deep-drawing process. J of Materials Proc. Tech., 125–126, pp. 440–445.

Michler, J.R., Weinmann, K.J., Kashani, A.R., Majlessi, S.A., (1994). A Strip drawing simulator with computer-controlled drawbead penetration and blankholder pressure. J. Materials Proc. Tech., Vol. 43, pp. 177–194.

Safonov, Michael G., (1980). Stability and Robustness of Multivariable Feedback Systems. MIT press.

Sheng, Z. G., Jirathearnat, S., Altan, T., (2004). Adaptive FEM Simulation for Prediction of Variable Blank Holder Force in Conical Cup Drawing. International J. of Machine Tools & Manuf., Vol. 44, pp. 487–494.

Siegert, K., Hohnhaus, J., Wagner, S., (1998). Combination of Hydraulic Multipoint Cushion System and Segment-Elastic Blankholders. SAE Paper No. 98007, pp. 51–55.
Siegert, K., Ziegler, M., Wagner, S., (1997). Loop Control of the Friction Force: Deep drawing process," J. of Materials Proc. Tech., Vol. 71, pp. 126–133.
Sklad, M.P., Harris, C.B., Slekirk, J.F., Grieshaber, D.J., (1992). Modelling of Die Tryout. SAE Paper No. 920433, pp. 151–157.
Sunseri, M., Cao, J., Karafillis, A.P., Boyce, M.C., (1996). Accommodation of Springback Error in Channel Forming Using Active Binder Force: Control Numerical Simulations and Experiments," J. of Engin. Materials and Tech., Vol. 118, pp. 426–435.
Wang, L., Lee, T.C., (2005). Controlled strain path forming process with space variant blank holder force using RSM method. J. of Materials Processing Tech., Vol. 167, pp. 447–455.
Yagami, T., Manabe, Ken-ichi, Yang, M., Koyama, H., (2004). Intelligent sheet stamping process using segment blank holder modules. J. of Materials Proc. Tech., Vol. 155–156: 2099–2105.
Ziegler, M., (1999) Pulsating Blankholder Technology. SAE Paper No. 1999-01-3155: 1–5.

Chapter 7
Auto-Tuning and Adaptive Control

Abstract This section describes the design and implementation of automatic controller tuning and model reference adaptive control (MRAC) to improve part quality in stamping and extends previous work on a manually-tuned fixed-gain process controller. Automatic tuning is described with a discussion of implementation issues in the presence of plant disturbances. Design of a direct MRAC, whose controller gains are continuously adjusted to accommodate changes in process dynamics and disturbances, is investigated, including simulation-based robustness analysis of the adaptation law and a consideration of constrained estimation in the recursive least squares algorithm to address practical implementation issues. The performance of the MRAC process controller designed through simulation is experimentally validated. Good tracking of the reference process variable (i.e., punch force), and significant part quality improvement in the presence of disturbances, is achieved.

7.1 Automatic Tuning and Adaptive Process Control

Sheet metal stamping is a widely-used manufacturing process for large-volume high-speed and low-cost production of body panels in automobiles, frame components in home-appliances, etc. In the previous chapter, we presented a multi-input multi-output (MIMO) process control system for stamping, which controls the punch force, F_p, at the four corners of the press, by adjusting the binder forces, F_b, using hydraulic actuators (Lim et al. 2010). The punch force is the total force applied by the drive mechanism of the press on the blank, and it is a function of the binder force and consequently, the restraining force on the blank. As described in Chap. 6 (see Fig. 6.1), punch force is affected by not only the binder forces, but also disturbance inputs. Thus, punch force is a suitable control variable for process control, as variations in part quality can be detected by measuring the punch force, and it can be controlled by adjusting the binder forces. The basis for using adjustable binder force is conceptually simple: if a split in the part is imminent, the force used

Y. Lim et al., *Process Control for Sheet-Metal Stamping*,
Advances in Industrial Control, DOI: 10.1007/978-1-4471-6284-1_7,

to hold the blank is reduced at the appropriate location, and allows more material flow to prevent the split. Conversely, if wrinkling is imminent in the part, the force holding the blank sheet is increased at the appropriate location on the blank, stretching out the wrinkle by restricting material flow (Cao and Boyce 1997; Sheng et al. 2004; Hsu et al. 2000, 2002; Doege et al. 2003; Lim et al. 2010; Siegert et al. 1997). Such a process controller enables the system to compensate for small variations in material thickness and properties in addition to lubrication changes, thus ensuring robustness in a typical production run with minimal operator intervention (Lim et al. 2010). As shown by the inner loop in Fig. 6.1, control of the binder force using hydraulic actuators is referred to as "machine" control, and is "open-loop" in terms of the process controlled variable (i.e., F_p). Thus, machine control based on fixed pre-determined desirable binder forces cannot maintain part quality in the presence of disturbances, which occur during production (Lim et al. 2010). Such disturbances can include change in material properties (e.g., formability, blank size, and sheet thickness), change in tooling (e.g., die wear), and variation in lubrication (Hardt 1993; Hsu et al. 2000; Yagami et al. 2004).

MIMO stamping process control has been designed using estimated MIMO process models based on system identification techniques, with manually-tuned fixed-gain proportional plus integral (PI) control, and has been shown to improve quality and consistency for a complex geometry part (i.e., an inner door panel for a pick-up truck) in the presence of plant disturbances (Lim et al. 2010). However, in those experimental tests manual controller fine-tuning, which can be time consuming and expensive, was required.

In this chapter, we consider two approaches to reduce the manual tuning effort for the MIMO process controller, namely, automatic controller tuning (or auto-tuning) and model reference adaptive control (MRAC). Industrial experience has clearly indicated that auto-tuning is a highly desirable and efficient method for PID controller tuning (Astrom et al. 1993; Astrom and Wittermark 1995). As shown in Fig. 7.1a, auto-tuning is a method to determine the process controller gains based on the observation that a process model has limit cycle oscillations with a specific period and amplitude in the output (y) under relay feedback. Controller gains are then calculated in terms of the period and amplitude of the output oscillation based upon an empirical rule-based table (i.e., Ziegler-Nichols ultimate stability method), which will be detailed in following section. However, if there are unpredictable parameters variations in the stamping process during operation, the fixed-gain PI process controller tuned by auto-tuning will still have limitations in tracking performance in the presence of such plant variations. Thus, we also consider an adaptive process controller whose parameters are continuously adjusted to accommodate changes in process dynamics and disturbances. The adaptive controller updates the gains of the MIMO PI process controller (which are initially set to the values obtained from auto-tuning) to minimize tracking error, in the presence of plant variations.

Adaptive control has been extensively studied during the last three decades for diverse applications (e.g., aircraft, automotive, and machine tools), e.g., (Rupp and Guzzella 2010; Boling et al. 2007; Tsai et al. 2009). Typically using a model reference control structure and normalized adaptive laws based upon a gradient or

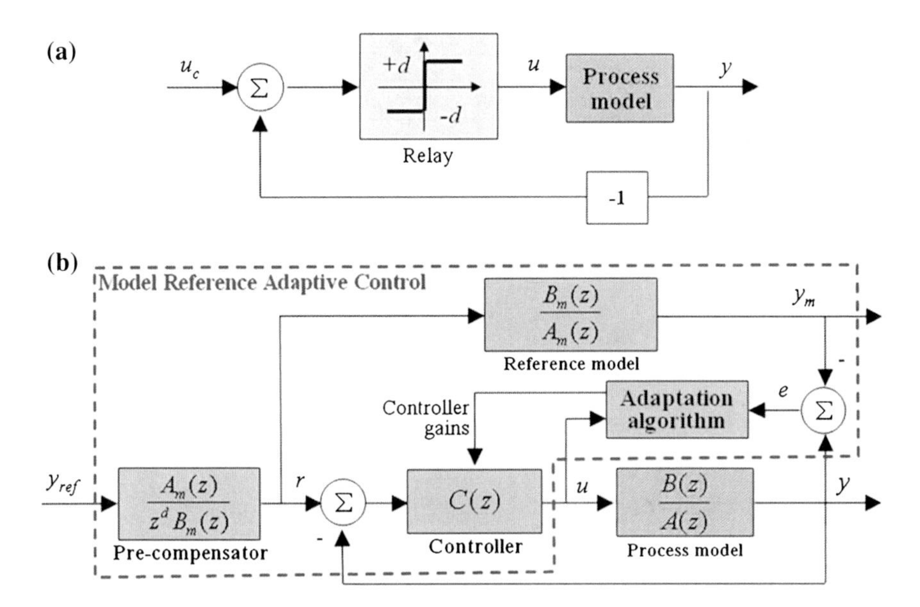

Fig. 7.1 Design methods for tuning, and fine-tuning process controllers: **a** auto-tuning with relay feedback **b** model reference adaptive control

recursive least squares on-line estimation (Lauderbaugh and Ulsoy 1989; Astrom and Wittermark 1995), the design and analysis of continuous- and discrete-time robust MRAC with normalized adaptive laws has been described in detail (Goodwin and Sin 1984; Ioannou and Sun 1996). Furthermore, in order to make adaptive control more attractive for practical implementation by effectively bounding control inputs, researchers have developed algorithms using constrained estimation (Goodwin and Sin 1984; Chia et al. 1991), and have performed sensitivity and robustness analyses (Ardalan and Adali 1989). In addition, in order to achieve high output tracking performance, pre-compensator or feedforward controllers based on the inverse of the closed-loop system or of the plant model have been studied (Tomizuka 1987; Devasia 2002; Karimi et al. 2008).

A discrete-time MRAC process controller is, for the first time, studied in stamping process control. It makes real-time adjustments to the binder force command based on feedback from the measurable process variable (i.e., punch force) in the presence of plant disturbances (e.g., lubrication and material thickness change). As shown in Fig. 7.1b, a direct MRAC structure includes a pre-compensator and minimizes the error between the reference model output (y_m) and the process model output (y). The reference model specifies the desired tracking performance of the closed-loop system, and the process model output represents the measured punch force in the presence of plant variations and disturbances.

Table 7.1 Estimated parameters of the 4th order perturbation model in the transfer function matrix with a sample rate of 100 Hz

$\delta F^1_{p,fil}$	$\delta F^2_{p,fil}$
$G^1_1(z) = \frac{0.8432\,Z^2 - 0.8321\,Z}{Z^4 - 0.9746\,Z^3 + 0.3065\,Z^2 - 0.3469\,Z + 0.1798}$	$G^2_1(z) = \frac{0.2462\,Z^2 - 0.1468\,Z}{Z^4 - 0.7203\,Z^3 - 0.1166\,Z^2 - 0.2076\,Z + 0.1942}$
$G^1_2(z) = \frac{2.095\,Z^2 - 1.586\,Z}{Z^4 - 0.9746\,Z^3 + 0.3065\,Z^2 - 0.3469\,Z + 0.1798}$	$G^2_2(z) = \frac{0.8538\,Z^2 - 0.7037\,Z}{Z^4 - 0.7203\,Z^3 - 0.1166\,Z^2 - 0.2076\,Z + 0.1942}$
$G^1_3(z) = \frac{1.526\,Z^2 - 1.479\,Z}{Z^4 - 0.9746\,Z^3 + 0.3065\,Z^2 - 0.3469\,Z + 0.1798}$	$G^2_3(z) = \frac{0.1353\,Z^2 - 0.0927\,Z}{Z^4 - 0.7203\,Z^3 - 0.1166\,Z^2 - 0.2076\,Z + 0.1942}$
$\delta F^3_{p,fil}$	$\delta F^4_{p,fil}$
$G^3_1(z) = \frac{0.4912\,Z^2 - 0.2313\,Z}{Z^4 - 1.403\,Z^3 + 0.3936\,Z^2 + 0.1452\,Z - 0.05513}$	$G^4_1(z) = \frac{0.7219\,Z^2 - 0.5878\,Z}{Z^4 - 1.621\,Z^3 + 0.6777\,Z^2 + 0.06901\,Z - 0.06487}$
$G^3_2(z) = \frac{0.333\,Z^2 - 0.03505\,Z}{Z^4 - 1.403\,Z^3 + 0.3936\,Z^2 + 0.1452\,Z - 0.05513}$	$G^4_2(z) = \frac{0.2685\,Z^2 - 0.1701\,Z}{Z^4 - 1.621\,Z^3 + 0.6777\,Z^2 + 0.06901\,Z - 0.06487}$
$G^3_3(z) = \frac{0.9474\,Z^2 - 0.66\,Z}{Z^4 - 1.403\,Z^3 + 0.3936\,Z^2 + 0.1452\,Z - 0.05513}$	$G^4_3(z) = \frac{0.4632\,Z^2 - 0.4214\,Z}{Z^4 - 1.621\,Z^3 + 0.6777\,Z^2 + 0.06901\,Z - 0.06487}$

7.2 Stamping Process Model and Estimation

In Chap. 6 and in our previous work (Lim 2010; Lim et al. 2010), we described the development of a MIMO linear sheet metal stamping process model for the purpose of controller design, taking into account the elongation of the sheet metal. The unknown parameters in the process models were estimated based on experimental data using the standard Least Squares (LS) algorithm. Based on data from system identification, the MIMO system is decomposed into four independent MISO system models, where each estimated model characterizes the dynamics of the process from the change of three reference binder force inputs to the change of one filtered punch force output. Moreover, the estimated parameters of the process models in discrete-time, which are shown in Table 7.1, have been validated by matching the experimentally measured punch force outputs with the punch force generated by the estimated models using the same commanded binder forces. The estimated process models are used as plant models in the auto-tuning procedure to obtain controller gains in Sect. 3 below. In addition, both estimated models and auto-tuned controller gains are used to compute the linear reference model, which specifies the desired performance of the closed-loop process, in the adaptive control design procedure.

7.3 Preliminary Experiment with Process Variables

As described in the previous section, the control-design model for stamping relates the binder force (i.e., input variable) and the punch force (i.e., output variable). The punch force varies with any change in the binder force. The punch force is also influenced by disturbances (e.g., lubrication and thickness change), under the same binder force conditions (Lim 2010; Lim et al. 2009, 2010). For example, as

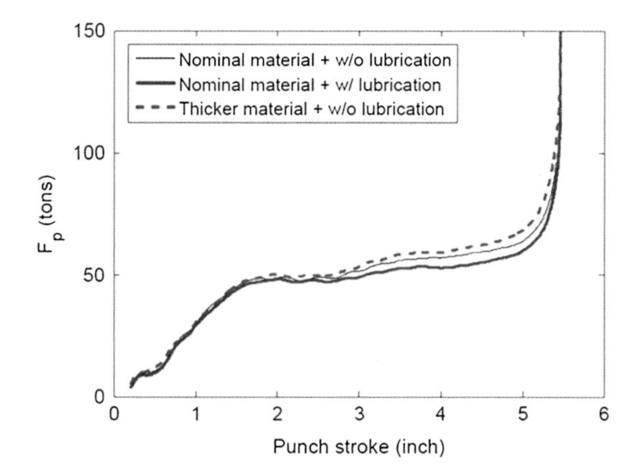

Fig. 7.2 The changes in the punch force due to plant disturbances (e.g., lubrication and thickness changes), with the same binder forces condition

illustrated in Fig. 7.2 for a representative test, when there is excessive lubricant on the sheet metal, the punch force output is smaller than under nominal conditions, with the same binder forces. Most blanks come with some form of lubrication; however, after a few thousand parts have been stamped, lubricant can build up in the die and cause wrinkling due to lower friction during the drawing process. Furthermore, the punch force also varies when a thicker (i.e., 0.79 mm) blank is used, as compared to a blank of nominal thickness (i.e., 0.64 mm), with the same binder forces and lubrication conditions. Material flow is restricted with a thicker blank, leading to splits. Figure 7.2 shows not only that more lubrication results in a lower punch force because of reduced friction/restraining force, but also that a thicker material results in a larger punch force because of increased friction/restraining force. We view these variations to be equivalent to binder-force disturbances, as the net effect is the same, and these variations can be corrected by binder force control. Thus, plant disturbances directly affect the material flow during the stamping process, leading to degradation of part quality. The effects of these disturbances are captured in the punch force data. Consequently, the punch force can be measured and used as a process variable to provide real-time feedback on part quality consistency in the presence of disturbances.

7.4 Auto-Tuning Based on Relay Feedback

7.4.1 Structure of the Process Controller

In our previous research (Lim 2010; Lim et al. 2009, 2010), as described in Chap. 6, four SIMO proportional plus integral (PI) controllers were implemented. The SIMO process controller at each corner in discrete-time is given by

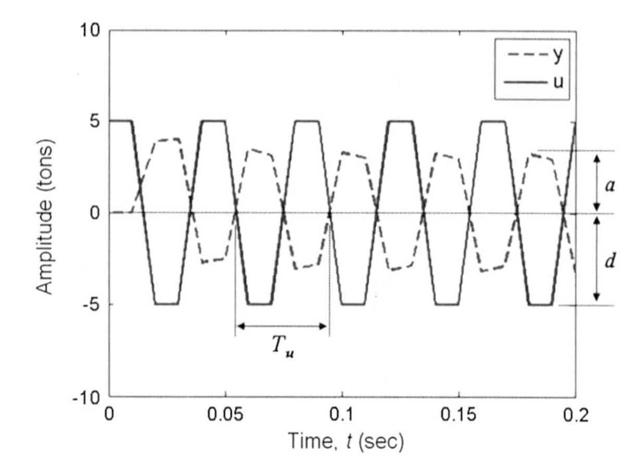

Fig. 7.3 Input and output oscillations of a process model with relay feedback

$$\underline{u}(k) = \left(\underline{K}_P + \underline{K}_I \frac{1}{z-1}\right)\{r(k) - y(k)\} \tag{7.1}$$

where $r(k) - y(k)$ is the error between the reference punch force ($F_{p,ref}$) and the measured punch force (F_p) at each corner, $\underline{K}_P$ is the vector of proportional control gains, and $\underline{K}_I$ is the vector of integral control gains. The Ziegler-Nichols rules (Ziegler and Nichols 1942) for a PI controller, as discussed in the next section, are used to obtain $\underline{K}_P$ and $\underline{K}_I$.

7.4.2 The Auto-Tuning Method

The basic idea for auto-tuning the process controller is the observation that many processes have limit cycle oscillations under relay feedback with step inputs (Astrom and Wittenmark 1995). As shown in Fig. 7.3, when the output lags behinds the input by $-\pi$ radians, the closed-loop system, which includes an ideal relay as shown in Fig. 7.1a, will oscillate with ultimate period (T_u) and amplitude (a). From the Fourier series expansion of the periodic relay output (u), the amplitude can be considered to be the result of the primary harmonic of the relay output. Therefore, the ultimate gain, K_u, can be obtained from the describing function approximation

$$K_u = \frac{4d}{\pi a} \tag{7.2}$$

where d is the magnitude of the relay and a is the amplitude of the output oscillation. Consequently, T_u and K_u can be used directly to obtain the controller gains. Based on the original Ziegler-Nichols tuning rules (Zielger and Nichols 1942), K_P and K_I, expressed in terms of K_u and T_u, are given by $K_P = 0.4 \times K_u$ and $K_I = K_P/(0.8 \times T_u)$.

In order to tune the PI gains numerically based on these rules, two parameters (i.e., a and T_u) of the output oscillation corresponding to a step command input (i.e., u_c), as shown Fig. 7.1a, are obtained through simulation, and the ultimate gain (K_u) is calculated using a (amplitude) and d (relay height) from Eq. 7.2. Three transfer functions, or MISO estimated process models for each corner (see Table 7.1), are used as the three different process models in Fig. 7.1a. Thus, three sets $\{(K_{P1}, K_{I1}), (K_{P2}, K_{I2}),$ and $(K_{P3}, K_{I3})\}$ of fixed gains of the SIMO PI process controller in Eq. (7.1) are simply tuned by three sets of T_u and K_u based on simulations performed by using three simultaneous inputs for the three transfer functions (or process models) for the punch force in each corner, ensuring that the auto-tuning accounts for the MISO structure. The main advantage of auto-tuning is that a single stamped part is needed to tune the controller, whereas without auto-tuning dozens of blanks may be wasted in establishing the appropriate gains for the PI controller. Simulation and experimental validation of this controller is described in following section, along with a comparison with the adaptive process controller that is described in the next section.

7.5 Design and Implementation of Direct MRAC

The direct model reference adaptive control (MRAC) approach dominates the adaptive control literature, due to the simplicity of its design as well as its robustness properties in the presence of process modeling errors (Astrom and Wittenmark 1995; Ioannou and Sun 1996). The basic structure of a direct MRAC is shown in Fig. 7.1b. The reference model is chosen to generate the desired performance of the closed-loop system, y_m, which the measured plant output, y (or the punch force), has to track. The tracking error, $e = y - y_m$, represents the deviation of the process model output from the desired punch force trajectory. In this section, the design and implementation of a direct MRAC process controller, which updates its controller gains to make the measured punch force (y) track the reference model output (y_m) as closely as possible in the presence of plant dynamics variations and disturbances, including a consideration of constraints in the recursive least squares (RLS) adaptation algorithm, is described.

7.5.1 *Direct MRAC Process Controller Structure*

The MISO estimated linear process model with the same denominator in discrete-time for each corner output) or y^i ($i = 1, 2, 3, 4$), is given as

$$y^i(k) = \frac{B_1^i(z)u_1^i(k)}{A^i(z)} + \frac{B_2^i(z)u_2^i(k)}{A^i(z)} + \frac{B_3^i(z)u_3^i(k)}{A^i(z)} \tag{7.3}$$

where

$$B_j^i(z) = b_{j1}^i z^2 + b_{j0}^i z \quad j = 1, 2, 3$$
$$A^i(z) = z^4 + a_3^i z^3 + a_2^i z^2 + a_1^i z + a_0^i$$

where $y^i = \delta F_{p,fil}^i$ and $\left[u_1^i \; u_2^i \; u_3^i\right]^T = \left[\delta F_{b1,ref}^i \; F_{b2,ref}^i \; F_{b3,ref}^i\right]^T$. All parameters of the perturbation process model were estimated using system identification as described in our previous work (Lim 2010; Lim et al. 2010).

Control Law The control law with a SIMO PI controller, and a pre-compensator to generate the reference input (r^i) is given as

$$u_j^i(k) = C_j^i(z)\left\{r^i(k) - y^i(k)\right\} \quad i = 1, 2, 3, 4 \tag{7.4}$$

where

$$C_j^i(z) = \left(\frac{s_{j1}^i z + s_{j0}^i}{z - 1}\right) \quad j = 1, 2, 3$$

and the SIMO PI controller parameters are related to the original K_P and K_I gains as:

$$s_{j1}^i = K_{Pj}^i \text{ and } s_{j0}^i = K_{Ij}^i - K_{Pj}^i, \quad j = 1, 2, 3 \tag{7.5}$$

The optimal values of these controller parameters are unknown, as the plant parameters vary from their nominal estimated values due to operational variation. Thus, the controller parameters s_{j1}^i and s_{j0}^i in Eq. (7.4) are replaced by their estimates $\hat{s}_{j1}^i(k)$ and $\hat{s}_{j0}^i(k)$ from the adaptation algorithm, to generate the control input as

$$u_j^i(k) = \left(\frac{\hat{s}_{j1}^i(k)z + \hat{s}_{j0}^i(k)}{z - 1}\right)\left\{r^i(k) - y^i(k)\right\} \quad j = 1, 2, 3 \tag{7.6}$$

Reference Model In the direct MRAC case, a reference model specifies the desired performance of the closed-loop system. In this study, as shown in Fig. 7.1b, the reference model in discrete-time was selected to be for each corner $i = 1, 2, 3, 4$:

$$\frac{y_m^i(k)}{r^i(k)} = \frac{B_m^i(z)}{A_m^i(z)} = \frac{b_{m3}^i z^3 + b_{m2}^i z^2 + b_{m1}^i z}{z^5 + a_{m4}^i z^4 + a_{m3}^i z^3 + a_{m2}^i z^2 + a_{m1}^i z + a_{m0}^i} \tag{7.7}$$

where this model has the same structure as the closed-loop system with a fixed-gain PI controller, and is obtained by combining Eq. (7.3) and(7.4), with the values of s_{j1}^i and s_{j0}^i set to their values obtained from auto-tuning, and by using the nominal process model parameters. The reference model based on the gains chosen using auto-tuning satisfies the step response specifications required for the stamping process, with a settling time of less than 0.1 s and an overshoot of less than 20 %.

Assumptions In order to meet the model reference control objective with an adaptive control law which is implementable, the plant model and the reference model need to satisfy the following assumptions (Ioannou and Sun 1996): First, the numerator $\left(B_j^i(z)\right)$ of the plant model as shown Eq. (7.3) must be a monic Hurwitz polynomial (i.e., minimum-phase). As shown in Table 7.1, all the estimated process models are minimum-phase systems based on a sample rate of 100 Hz. Second, A_m^i and B_m^i of the reference model must be also monic Hurwitz polynomials. In other words, the locations of poles and zeros of the reference models are inside the unit circle in the z-plane. Finally, the relative degree (i.e., 2) of the reference model shown in Eq. (7.7) must be the same as the relative degree (i.e., 2) of the plant model in Eq. (7.3).

Adaptive Law The adaptive law for generating $\hat{s}_{j1}^i(k)$ and $\hat{s}_{j0}^i(k)$ at time-step k is developed by viewing the problem as an on-line estimation problem for s_{j1}^i and s_{j0}^i. This is accomplished by obtaining an appropriate parameterization for the MISO process model in Eq. (7.3) for on-line estimation, in terms of the unknown parameters s_{j1}^i and s_{j0}^i.

The MISO plant model for each corner ($i = 1, 2, 3, 4$) is given in Eq. (7.3). Also, the control law, for the SIMO PI controller, is given in Eqs. (7.4) and (7.5). Adding and subtracting $u_j^i(k)$ from Eq. (7.4) into Eq. (7.3) yields

$$\begin{aligned} y^i &= \frac{B_1^i(C_1^i - C_1^i) + B_2^i(C_2^i - C_2^i) + B_3^i(C_3^i - C_3^i)}{A^i}(r^i - y^i) + \frac{B_1^i u_1^i + B_2^i u_2^i + B_3^i u_3^i}{A^i} \\ &= \frac{B_1^i C_1^i + B_2^i C_2^i + B_3^i C_3^i}{A^i}(r^i - y^i) - \frac{B_1^i C_1^i + B_2^i C_2^i + B_3^i C_3^i}{A^i}(r^i - y^i) + \frac{B_1^i u_1^i + B_2^i u_2^i + B_3^i u_3^i}{A^i} \end{aligned} \tag{7.8}$$

Multiplying Eq. (7.8) on both sides by $\frac{A^i}{A^i + B_1^i C_1^i + B_2^i C_2^i + B_3^i C_3^i}$, it follows that

$$\begin{aligned} y^i &= \frac{B_1^i C_1^i + B_2^i C_2^i + B_3^i C_3^i}{A^i + B_1^i C_1^i + B_2^i C_2^i + B_3^i C_3^i} r^i \\ &\quad - \frac{B_1^i C_1^i + B_2^i C_2^i + B_3^i C_3^i}{A^i + B_1^i C_1^i + B_2^i C_2^i + B_3^i C_3^i}(r^i - y^i) + \frac{B_1^i u_1^i + B_2^i u_2^i + B_3^i u_3^i}{A^i + B_1^i C_1^i + B_2^i C_2^i + B_3^i C_3^i} \end{aligned} \tag{7.9}$$

Based on Fig. 7.1b, since $A^i(z) \in R^1$, $B_j^i(z) \in R^{1\times3}$ and $C_j^i(z) \in R^{3\times1}$ are defined for each corner i *as* shown in Eq. (7.3) and (7.4) respectively, the closed-loop transfer function is given by

$$\begin{aligned}\frac{y^i(k)}{r^i(k)} &= \frac{\frac{B_1^i C_1^i + B_2^i C_2^i + B_3^i C_3^i}{A^i}}{1 + \frac{B_1^i C_1^i + B_2^i C_2^i + B_3^i C_3^i}{A^i}} \\ &= \frac{B_1^i C_1^i + B_2^i C_2^i + B_3^i C_3^i}{A^i + B_1^i C_1^i + B_2^i C_2^i + B_3^i C_3^i}\end{aligned} \tag{7.10}$$

Hence, the reference model which specifies the desired performance of the closed-loop system has the form shown in Eq. (7.10), i.e.,

$$\begin{aligned}\frac{y^i}{r^i} &= \frac{B_1^i C_1^i + B_2^i C_2^i + B_3^i C_3^i}{A^i + B_1^i C_1^i + B_2^i C_2^i + B_3^i C_3^i} \\ &\triangleq \frac{y_m^i}{r^i} = \frac{B_m^i}{A_m^i}\end{aligned} \tag{7.11}$$

and is obtained by using the controller parameters from auto-tuning in C_j^i, and the nominal estimated plant parameters in B_j^i and A^i. Thus, defining $A_m^i = A^i + B_1^i C_1^i + B_2^i C_2^i + B_3^i C_3^i$ and $B_m^i = B_1^i C_1^i + B_2^i C_2^i + B_3^i C_3^i$ in Eq. (7.11), then Eq. (7.9) can be written as

$$y^i = \frac{B_m^i}{A_m^i} r^i - \frac{B_1^i C_1^i + B_2^i C_2^i + B_3^i C_3^i}{A_m^i}\left(r^i - y^i\right) + \frac{B_1^i u_1^i + B_2^i u_2^i + B_3^i u_3^i}{A_m^i} \tag{7.12}$$

or

$$e^i = -\frac{B_1^i C_1^i + B_2^i C_2^i + B_3^i C_3^i}{A_m^i}\left(r^i - y^i\right) + \frac{B_1^i u_1^i + B_2^i u_2^i + B_3^i u_3^i}{A_m^i} \tag{7.13}$$

with $y_m^i = \frac{B_m^i}{A_m^i} r^i$, as given in Eq. (7.11). Then tracking error is defined as $e^i = y^i - y_m^i$. If $C_1^i, C_2^i,$ and C_3^i as given previously in Eq. (7.4), and then replaced by the estimated controller gains ($\hat{s}_{j1}^i$ and $\hat{s}_{j0}^i$), are substituted into Eq. (7.13)

$$\begin{aligned}e^i = &-\frac{B_1^i}{A_m^i}\left(\frac{\hat{s}_{11}^i z + \hat{s}_{10}^i}{z-1}\right)\left(r^i - y^i\right) - \frac{B_2^i}{A_m^i}\left(\frac{\hat{s}_{21}^i z + \hat{s}_{20}^i}{z-1}\right)\left(r^i - y^i\right) - \frac{B_3^i}{A_m^i}\left(\frac{\hat{s}_{31}^i z + \hat{s}_{30}^i}{z-1}\right)\left(r^i - y^i\right) \\ &+ \frac{B_1^i u_1^i + B_2^i u_2^i + B_3^i u_3^i}{A_m^i}\end{aligned} \tag{7.14}$$

Thus, Eq. (7.14) can be written in the parametric model form for adaptive updates in the direct MRAC as

$$e^i = \left(\theta^i \phi^{iT} + u_f^i\right) \quad i = 1, 2, 3, 4 \tag{7.15}$$

where

$$\theta^i = \left[\hat{s}^i_{11}, \hat{s}^i_{10}, \hat{s}^i_{21}, \hat{s}^i_{20}, \hat{s}^i_{31}, \hat{s}^i_{30}\right]$$

$$\phi^{iT} = \begin{bmatrix} -\frac{z}{z-1}\frac{B^i_1}{A^i_m} \\ -\frac{1}{z-1}\frac{B^i_1}{A^i_m} \\ -\frac{z}{z-1}\frac{B^i_2}{A^i_m} \\ -\frac{1}{z-1}\frac{B^i_2}{A^i_m} \\ -\frac{z}{z-1}\frac{B^i_3}{A^i_m} \\ -\frac{1}{z-1}\frac{B^i_3}{A^i_m} \end{bmatrix} \left(r^i - y^i\right)$$

$$u^i_f = \frac{B^i_1 u^i_1 + B^i_2 u^i_2 + B^i_3 u^i_3}{A^i_m}$$

where B^i_j and A^i_m are given in Eqs. (7.3) and (7.7) respectively.

The adaptation algorithm, based on the parametric model above, is based on the RLS algorithm with exponential forgetting (Astrom and Wittenmark 1995). Assuming that the matrix ϕ^i has full rank, the controller parameters, θ^i, are estimated recursively using the update law:

$$\theta^i(k) = \theta^i(k-1) + P^i(k)\phi^i(k) \underbrace{\left\{e^i(k) - \left(\phi^{iT}(k)\theta^i(k-1) + u^i_f(k-1)\right)\right\}}_{\varepsilon^i(k)}$$

$$P^i(k) = \frac{1}{\lambda}\left[P^i(k-1) - \frac{P^i(k-1)\phi^i(k)\phi^{iT}(k)P^i(k-1)}{\lambda I + \phi^{iT}(k)P^i(k-1)\phi^i(k)}\right] \tag{7.16}$$

where $\varepsilon^i(k)$ is the estimation error and $e^i(k) = y^i(k) - y^i_m(k)$ is the tracking error for each corner i (i = 1, 2, 3, 4). The forgetting factor, $0 < \lambda \leq 1$, has the interpretation that if $\lambda = 1$ the algorithm reduces to the standard RLS algorithm and as λ gets smaller the algorithm "discards" older data more quickly. Moreover, one of the important tuning factors in the standard RLS algorithm is the initialization of the updating covariance matrix (i.e., $P^i(k)$). If the initial values of this covariance matrix are large, the controller gain updates can lead to large transients and a phenomenon referred to as bursting. Thus, the initial values of the covariance matrix are set to small values to start with, and then increased, if required, to achieve better performance during implementation.

Pre-Compensator For improved tracking performance, the inverse dynamics of the reference model shown in Eq. (7.7) are utilized in the pre-compensator (see Fig. 7.1b), and then delayed by two time steps, in order to make the pre-compensator causal. Thus, the transfer function of the pre-compensator, or $G_{pc}(z)$, is expressed as

$$G_{pc}(z) = \frac{r^i(k)}{y^i_{ref}(k)} = \frac{1}{z^d}\frac{A^i_m(z)}{B^i_m(z)} \tag{7.17}$$

where z^d represents a d-step delay (i.e., $d = 2$ here) used to make the pre-compensator causal. As described previously in the assumptions, $B^i_m(z)$ is a monic Hurwitz polynomial which guarantees the stability of the pre-compensator.

In addition, note that the estimated plant model and the auto-tuned gains are used for two purposes in the MRAC scheme: (1) to initialize the gains of the adaptive controller and generate the required regressor vectors, and (2) to define the reference model. Clearly, if the actual plant is exactly the same as the estimated model, the reference model and plant outputs will be equal and the controller parameters will not be updated, resulting in closed-loop performance identical to that of the reference model.

7.5.2 Robustness of MRAC to Parameters Variations

The robustness of the direct MRAC to process model perturbations, which represent unexpected plant dynamics changes and/or a slowly time-varying system, is considered using a simulation-based approach. The RLS adaptation algorithm, which is designed using both nominal plant model coefficients and auto-tuned PI gains, is also a critical part of the MRAC controller that depends on the plant model. Thus, in order to analyze robustness of the RLS adaptation algorithm with respect to plant variations, the estimated process model coefficients in simulation are perturbed to study the resulting closed-loop performance.

The performance of the RLS algorithm can be analyzed by considering tracking error and estimation error variations caused by unexpected process dynamics changes. Equation (7.16) can be rewritten using prime notation to denote plant variations as

$$\theta^{i'}(k) = \theta^{i'}(k-1) + P^{i'}(k)\phi^{i'}(k)\underbrace{\left\{e^{i'}(k) - \left(\phi^{iT'}(k)\theta^{i'}(k-1) + u^{i'}_f(k-1)\right)\right\}}_{\varepsilon^{i'}(k)} \tag{7.18}$$

and the tracking error variation can be given as

$$\delta e^i(k) = e^i(k) - e^{i'}(k) \tag{7.19}$$

where $e^i(k)$ is the tracking error without plant variation, and $e^{i'}(k)$ is the tracking error with plant variations. Also, the estimation error variation can be obtained as

$$\delta\varepsilon^i(k) = \varepsilon^i(k) - \varepsilon^{i'}(k) \tag{7.20}$$

where $\varepsilon^i(k)$ is the estimation error without plant variation, and $\varepsilon^{i'}(k)$ is the estimation error with plant variations.

For robustness analysis of the RLS algorithm, randomly chosen process models, which vary within ±30 % of each estimated process model parameter shown in Table 7.1, are used. Simulation results, not shown here, showed good tracking performance with relatively large plant parameter variations of 30 % is achieved with small tracking errors (i.e., less than 5 % of the punch force output) (Lim 2010).

7.5.3 Parameter-Constrained Estimation

For systems that are time-varying or nonlinear, constrained parameter estimation, based on experimental data or other a priori knowledge, can reduce or eliminate problems, such as temporary estimation gain bursting, large transients and offset due to disturbances (Goodwin and Sin 1984; Chia et al. 1991). Instead of using *unconstrained* estimates for θ^i in Eq. (7.16) the constraints on individual controller parameters (i.e., adaptive controller gain bounds) can be utilized as follows (Goodwin and Sin 1984):

$$\theta^i_{n,c} = \begin{cases} \theta^i_{n,\min} & \text{if} \quad \theta^i_n < \theta^i_{n,\min} \\ \theta^i_n & \text{if} \quad \theta^i_{n,\min} \leq \theta^i_n \leq \theta^i_{n,\max} \\ \theta^i_{n,\max} & \text{if} \quad \theta^i_n > \theta^i_{n,\max} \end{cases} \quad n = 1, \ldots, 6 \tag{7.21}$$

where the subscript c specifies that θ^i_n is the *constrained* solution for each of the controller parameters denoted by n (i.e., $n = 1, 2, 3, \ldots, 6$) in each corner i. As shown in Fig. 7.4, the PI controller gains estimated using the constrained algorithm are bounded, while the controller gains updated using the unconstrained algorithm show large transients in simulation.

Therefore, as shown in Fig. 7.5, the constrained estimation algorithm not only enables the punch force output to achieve better tracking of the measured reference punch force than the unconstrained estimates, but also ensures that no mechanical damage is caused by over-driving the actuators. In this simulation the auto-tuned PI gains described previously are used as initial values of the process controller gains in the RLS algorithm, and estimated process models are used as the process model. Initial values of the covariance matrix P^i (see Eq. (7.16)) are set to $10^{-2} \times I(n, n)$ where I is the identity matrix and n (i.e., $n = 6$) is the number of controller gains for each corner output. The forgetting factor (λ) in the RLS algorithm is set to one resulting in a standard implementation. The constraints on the parameters are established via simulation to ensure that the commands to the actuators are within their saturation limits.

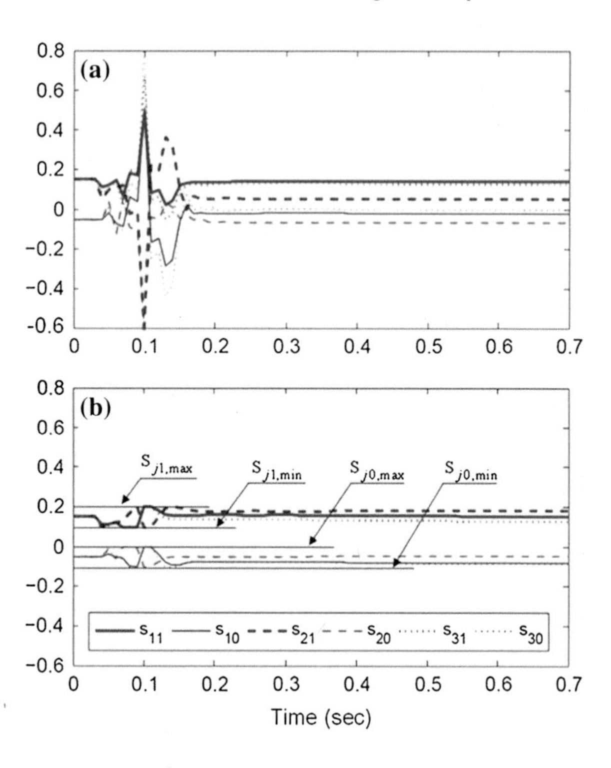

Fig. 7.4 Comparison of estimated controller gains using the RLS algorithm: **a** unconstrained **b** constrained

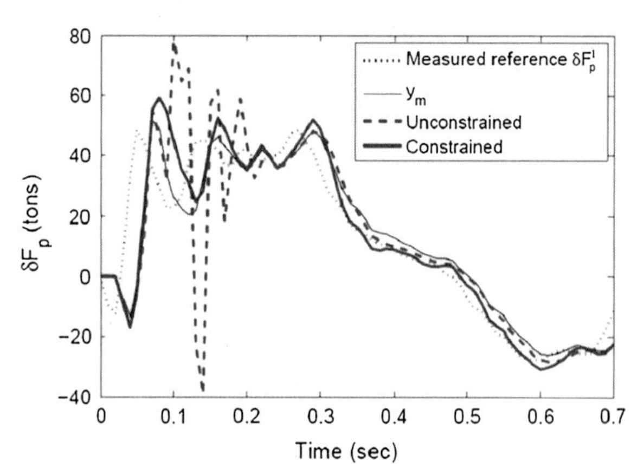

Fig. 7.5 Simulation results for punch force output tracking performance with two estimation algorithms: unconstrained and constrained

7.6 Simulation Results

Simulation is used to validate the performance of the two proposed PI process controllers (i.e., auto-tuned and direct MRAC process controller) based on the estimated perturbation process models, with randomly-assigned process model parameter variations to represent changes in plant dynamics. The simulation

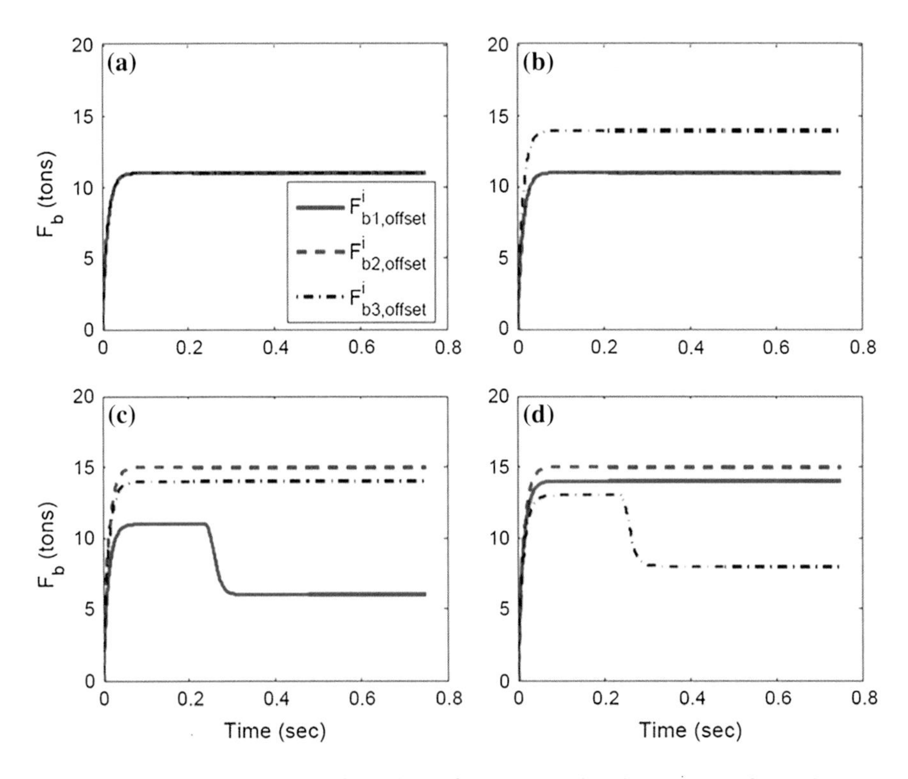

Fig. 7.6 The pre-determined nominal binder force trajectories (i.e., $F^i_{bj,offset}$) for each corner output i: **a** $i = 1$ **b** $i = 2$ **c** $i = 3$ **d** $i = 4$

models use the perturbed binder forces $\left(\delta F^i_{bj}\right)$ as inputs, the perturbed punch force $\left(\delta F^i_p\right)$ as output, and the desired perturbed punch force $\left(\delta F^i_{p,ref}\right)$ as the reference (or desired punch force) for each corner i.

The three binder forces $\left(F^i_{bj}, j = 1, 2, 3\right)$ associated with each punch force corner, i, ($i = 1, 2, 3, 4$) are updated to minimize the difference between F^i_p and $F^i_{p,ref}$, by adding or subtracting the perturbed binder forces $\left(\delta F^i_{bj}\right)$, which are produced by the process controller. Thus, the total binder forces, F^i_{bj}, in both simulation and experiment are given by

$$F^i_{bj} = F^i_{bj,offset} + \delta F^i_{bj} \quad j = 1, 2, 3 \tag{7.22}$$

where $F^i_{bj,offset}$ are pre-determined nominal binder forces for each corner, i, and are shown in Fig. 7.6. As described previously, the machine control (MC) alone (without the process controller) generates these pre-determined nominal binder forces, even in the presence of disturbances.

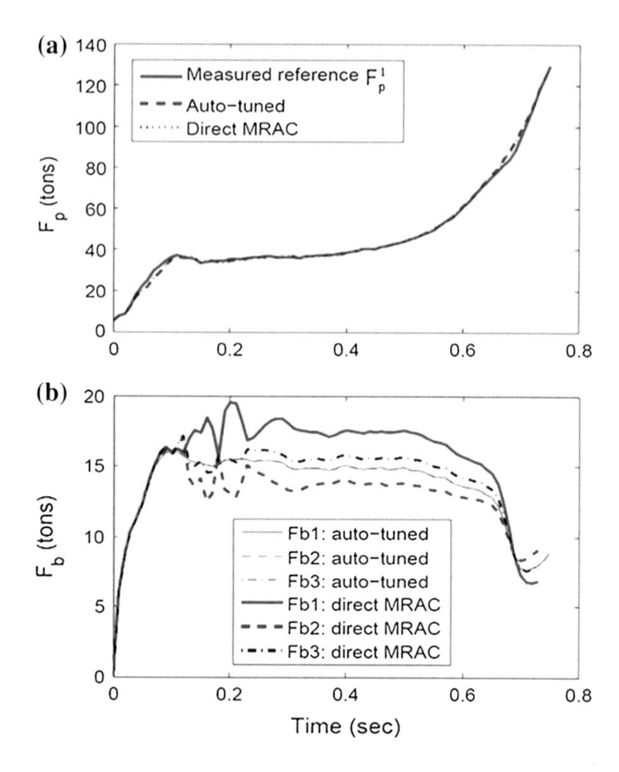

Fig. 7.7 Simulation results comparing two process controllers (i.e., the auto-tuned fixed PI controller and the direct MRAC): **a** punch force (i.e., F_p^1) **b** binder forces associated with the punch force

The reference punch force trajectories $\left(F^i_{p,ref}\right)$ are obtained by recording the punch forces generated using the pre-determined nominal binder forces which are determined by experienced die-makers to make a good part using the material with nominal properties and under normal operating conditions for each punch force corner i.

Figure 7.7 shows the simulation results with the punch force as output and the binder force as input, using the fixed-gain auto-tuned PI process controller and the direct MRAC PI process controller based on the process models, which include 30 % changes in the parameters as plant variations. In Fig. 7.7, when the controlled punch force becomes smaller than the reference punch force (e.g., at the beginning and the end of the stroke), both PI process controllers enable the punch force to track the reference punch force by adjusting the binder forces.

The performance of the two controllers in the presence of actual plant variations and disturbances in the form of intentionally introduced lubrication and material thickness changes is described in the next section on experimental validation.

7.7 Experimental Validation

The two MIMO PI process controllers (i.e., fixed-gain PI obtained by auto-tuning and direct MRAC PI process controller) described above were implemented on the experimental system described in (Lim et al. 2009, 2010). Their performance was compared to the performance of the machine control (MC) only (i.e., without process control (PC)) with fixed pre-determined binder forces commands ($F^{i}_{bj,offset}$ in Eq. (7.22)) in terms of deviation from the reference punch force, and in terms of part quality, in the presence of disturbances.

7.7.1 Lubrication Change

The first disturbance that we consider is lubrication change. Figure 7.8 shows the tracking performance of the punch force for the test cases, and illustrates stamped part quality comparison for those cases in the presence of lubrication change. Based on convention in the stamping industry to view the forming process in terms of punch stroke to eliminate variability in terms of press speeds, the punch force data are plotted with respect to punch stroke on the horizontal axis. Due to space limitations results are presented for only one corner of the punch forces (i.e., F^{1}_{p}). However, all corners of the punch forces and part quality for all corners were investigated and showed similar trends. Figure 7.8a shows that with only machine control, (i.e., without process control) there is a clear difference between the reference punch force which characterizes a good part in the absence of the extra lubrication and the measured punch force in the presence of the extra lubrication. Figure 7.8c shows that extra lubrication results in significant wrinkling, caused by greater material flow. The desired part quality, characterized by the reference punch force, is shown in Fig. 7.8b. In Fig. 7.8a, d, and e, it can be seen that the MRAC process controller provides the best tracking performance of the reference trajectory as well as part quality improvement in the presence of lubrication change. The fixed-gain auto-tuned PI process controller also enables the punch force to track the reference punch force in an average sense, but with some oscillation throughout the punch stroke under the excessive lubrication condition. It is noted that these oscillations may be attenuated by further manual tuning as per results in previous work (Lim et al. 2009, 2010); however, our objective is to automate the tuning effort as shop-floor personnel may not have the expertise and time to do so. In spite of the oscillations, clear improvement in part quality is observed. Thus, the above experimental results, using a complex geometry part, show that the MIMO process controller, designed through simulation, is quite effective in improving part quality in the presence of lubrication change.

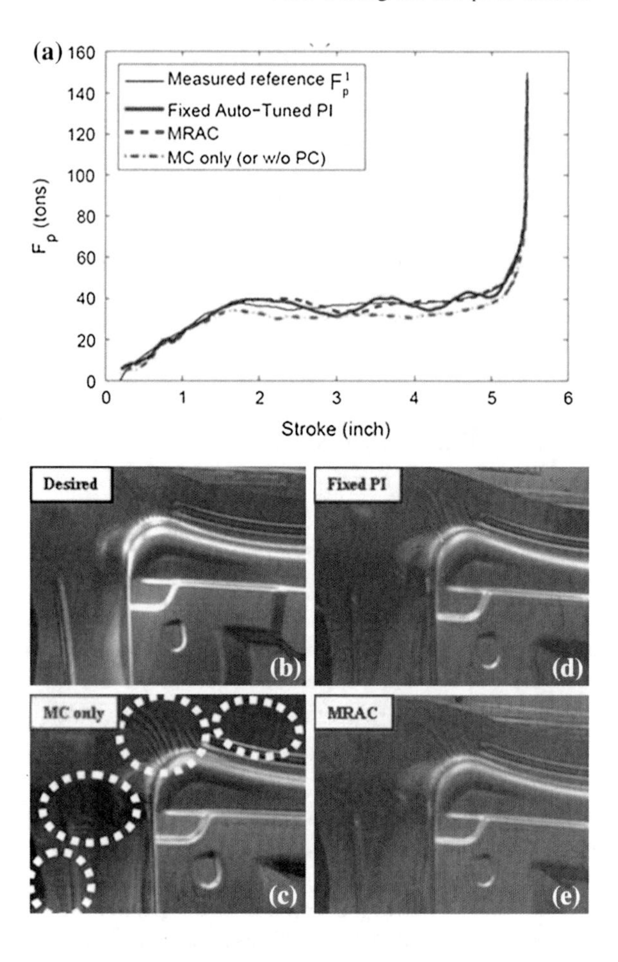

Fig. 7.8 Experimental results in the presence of lubrication change: **a** the punch force (i.e., F_p^1) **b** dry condition and **c–e** excessive lubricated condition

7.7.2 Material Thickness Variation

The second disturbance that we consider is material thickness variation. With thicker material (i.e., 0.79 mm) compared to the nominal (i.e., 0.64 mm), Fig. 7.9 shows that the two process controllers effectively track the reference punch force, which was determined using the nominal material. We intentionally introduce a relatively large change in thickness to show that process control compensates for such large variations and would, thus, be effective for the smaller variations typically seen in production. Again, as shown in Fig. 7.9, machine control alone (i.e., without process control), which has fixed pre-determined desirable binder forces, cannot minimize the error between the reference punch force and the measured punch force. However, the direct MRAC process controller shows good tracking performance of the measured reference punch force by adjusting the three associated binder forces (see Fig. 7.9b) from their nominal values (i.e., $F^i_{bj,offset}$). For example, when the measured punch force is greater than the reference punch

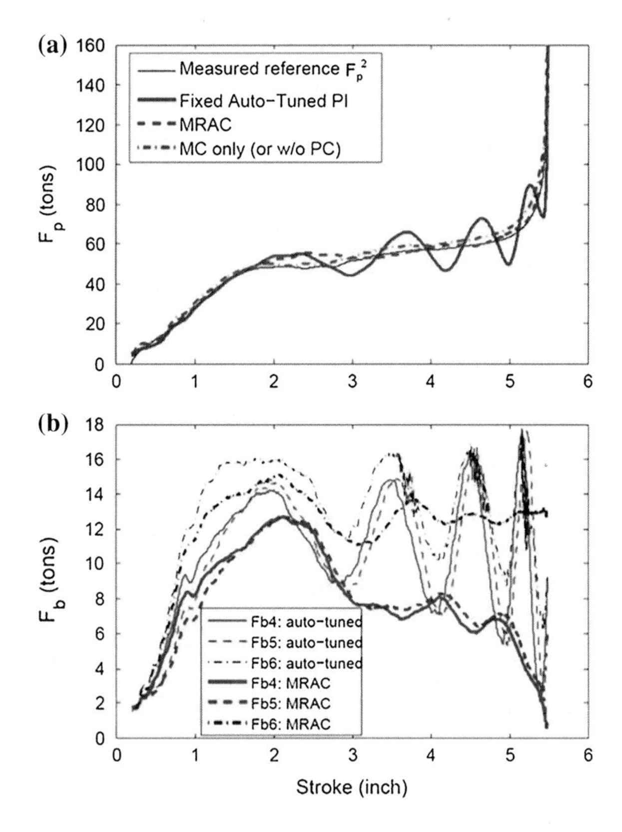

Fig. 7.9 Experimental results in the presence of material thickness variation: **a** punch force (i.e., F_p^2) **b** binder forces

force, at around 2.5 inches of punch stroke, the MRAC process controller drastically reduces the binder forces in order to minimize the difference between the reference F_p^2 and measuredF_p^2. The fixed-gain auto-tuned PI controller performs less effectively with material thickness change, and shows excessive oscillations.

7.8 Concluding Remarks

The results presented show that a "machine control only" strategy can be improved upon using process control, in the presence of disturbances, in terms of both tracking performance and part quality. In this paper we introduce auto-tuning and direct MRAC to minimize the manual-tuning effort for the MIMO stamping process control. A simulation-based auto-tuning method, which requires the stamping of only one additional part for the purpose of plant model parameter estimation, has been investigated to eliminate a manual tuning approach to obtain fixed PI gains. Furthermore, auto-tuning provides an effective way to initialize the design and implementation of an adaptive stamping process controller, based on the MRAC approach. The direct MRAC stamping process controller works much better than the fixed PI process controller based on auto-tuning; however, the auto-tuning method is useful for providing good initial gain values for use with the adaptive controller.

In this chapter we have described the design and implementation of a novel MIMO direct MRAC stamping process controller, and experimentally validated the tracking performance of the reference punch force as well as part quality improvement, in the presence of plant variations and/or disturbances. The direct MRAC stamping process controller, which includes a pre-compensator, provided excellent tracking performance, and resulted in good part quality, even in the presence of significant disturbances (i.e., dry versus lubricated or almost 25 % increase in thickness). The use of the constrained RLS algorithm for the MRAC yielded better adaptation results, eliminating problems related to large transients. Finally, the results presented show that use of nominal parameter values based on system identification in the pre-compensator, and for the filter in the regressor vector for MRAC, is effective.

The direct MRAC stamping process controller can benefit from simplification of the design procedures for more user-friendly implementation in a production environment. Thus, we are currently investigating the feasibility of removing the system identification and auto-tuning steps from the adaptive control implementation.

References

Ardalan, S.H., Adali, T., (1989). Sensitivity Analysis of Transversal RLS Algorithms with Correlated Inputs. *IEEE Inter. Symposium on Circuits and Systems*, Vol. 3, pp. 1744–1747.

Astrom, K.J., Hagglund, T., Hang, C.C., Ho, W.K., (1993). Automatic Tuning and Adaptation for PID Controller. *Control Engineering and Practice*, Vol. 1, No. 4, pp. 699–714.

Astrom, K.J. and Wittenmark, B., (1995). *Adaptive Control*. Prentice Hall, 2nd edition.

Boling, J.M., Seborg, D.E., Hespanha, J.P., (2007). Multi-model adaptive control of a simulated pH neutralization process. *Control Engineering Practice*, Vol. 15, Issue 6, Special Section on Control Applications in Marine Systems—CAMS2004, Control Applications in Marine Systems, June 2007, pp. 663–672.

Cao, J. and Boyce, M.C., (1997). A Predictive Tool for Delaying Wrinkling and Tearing Failure in Sheet Metal Forming. *J. of Eng. Materials and Technology*, Vol. 119.

Chia, T.L., Chow, P.-C., Chizeck, H.J., (1991). Recursive Parameter Identification of Constrained Systems: An Application to Electrically Stimulated Muscle. *IEEE Trans. Biomed. Eng.*, 38, pp. 429–442.

Devasia, S., (2002). Should Model-Based Inverse Inputs Be Used as Feedforward Under Plant Uncertainty?. *IEEE Trans. on Automatic Control*, Vol. 47, No. 11, pp. 1865–1871.

Doege, E., Schmidt-Jurgensen, R., Huinink, S., Yun, J.-W, (2003). Development of an optical sensor for the measurement of the material flow in deep drawing processes. *CIRP Annals—Manufacturing Technology*, Vol 52, n1, pp. 225–228.

Goodwin, G.C. and Sin, K.W., (1984). *Adaptive Filtering, Prediction, and Control*. Prentice-Hall: Englewood Cliffs, NJ., pp. 91–94.

Hardt, D.E., (1993). Modeling and Control of Manufacturing Processes: Getting More Involved. *ASME Journal of Dyn. Systems, Meas., and Control*, Vol. 115, pp. 291–300.

Hsu, C.W., Ulsoy, A.G., Demeri, M.Y., (2000). An Approach for Modeling Sheet Metal Forming for Process Controller Design. *ASME J. Manuf. Sci. Eng.*, 122, pp. 717–724.

Hsu, C.W., Ulsoy, A.G., Demeri, M.Y., (2002). "Development of process control in sheet metal forming," *J. of Materials Proc. Tech.*, Vol. 127, pp. 361–368.

Ioannou, P.A. and Sun, J., (1996). *Robust Adaptive Control*. New Jersey: Prentice-Hall.

Karimi, A., Butcher, M., Longchamp, R., (2008). Model-Free Pre-compensator Tuning Based on the Correlation Approach. *IEEE Trans. on Control Systems Tech.*, Vol. 16, pp. 1013–1020.

Lauderbaugh, L.K. and Ulsoy, A.G., (1989). Model Reference Adaptive Control in Milling. *ASME J. of Eng. for Ind.*, Vol. 111, pp. 13–21.

Lim, Y.S., (2010). *MIMO Adaptive Process Control in Stamping Using Punch Force*. PhD Dissertation, University of Michigan, Ann Arbor, Michigan.

Lim, Y.S., Venugopal, R., Ulsoy, A.G., (2009). Improved Part Quality in Stamping Using Multi-Input Multi-Output Process Control. *Proceedings of American Control Conference*, Saint Louis, MO, USA, June 10–12, 2009, pp. 5570–5575.

Lim, Y.S., Venugopal, R., Ulsoy, A.G., (2010). Multi-Input Multi-Output Modeling and Control for Stamping. *ASME J. of Dyn. Systems, Meas., and Control*, Vol. 132, Issue 4, 041004 (12 pages).

Rupp, D. and Guzzella, L., (2010). Adaptive internal model control with application to fueling control. *Control Engineering Practice*, Vol. 18, Issue 8, pp. 873–881.

Sheng, Z., Jirathearanat, S., Altan, T., (2004). Adaptive FEM Simulation for Prediction of Variable Blank Holder Force in Conical Cup Drawing. *International Journal of Machine Tool and Manufacture*, 44, pp. 487–494.

Siegert, K., Ziegler, M., Wagner, S., (1997). Loop Control of the Friction Force: Deep drawing process. *J. of Materials Proc. Tech.*, Vol. 71, pp. 126–133.

Tomizuka, M., (1987). Zero Phase Error Tracking Algorithm for Digital Control. *ASME J. of Dynamic Systems, Measurement, and Control*, Vol. 109, pp. 65–68.

Tsai, Ching-Chih, Lin, Shui-Chun, Wang, Tai-Yu, Teng, Fei-Jen, (2009). Stochastic model reference predictive temperature control with integral action for an industrial oil-cooling process. *Control Engineering Practice*, Vol. 17, Issue 2, pp. 302–310.

Yagami, T., Manabe, K., Yang, M., Koyama, H., (2004). Intelligent Sheet Stamping Process Using Segment Blank Holder Modules. *J. of Mat. Proc. Tech.*, Vol. 155–156, pp. 2099–2105.

Ziegler, J.G. and Nichols, G.A., (1942). Optimum Settings for Automatic Controllers, *Trans. ASME* 64, pp. 759–768.

Chapter 8
Direct and Indirect Adaptive Process Control

Abstract This chapter compares the design, implementation and performance of direct and indirect adaptive control (AC) to improve part quality in the stamping process in the presence of disturbances. First, previous work on the design and performance of a direct AC approach (i.e., model reference adaptive control or MRAC) is summarized. The direct AC filter uses nominal process parameters, and so requires some knowledge of the process. Consequently, an indirect AC approach, which estimates process parameters on-line, was also considered. However, due to the simple proportional plus integral (PI) control structure selected, the computation of the controller gains from the estimated parameters requires an optimization procedure, which is not amenable to real-time implementation. Thus, the indirect AC is implemented using a look-up table, with controller gains that are pre-computed off-line via optimization. The indirect AC with the look-up table is compared to the direct AC via simulations and experiments in terms of tracking performance as well as part quality, in the presence of plant variations. The indirect AC, with a sufficiently high level of discretization in the look-up table, performs well in simulations. However, due to extensive memory requirements, a smaller look-up table is used in the experiments, where it is outperformed by the direct AC.

8.1 Adaptive Control Applications

Adaptive control has been extensively studied during the last 3 decades for diverse applications (e.g., aircraft, automotive, process control and machine tools), see, for example, (Rupp and Guzzella 2010; Boling et al. 2007; Tsai et al. 2009; Lauderbaugh and Ulsoy 1989). An adaptive controller (AC) is defined as a controller with adjustable control parameters (or gains) and an adaptation law for adjusting the control parameters to achieve a desired control objective. In indirect AC the process parameters are estimated on-line and used to calculate the controller gains. However, in direct AC the controller gains are directly adjusted (or

Y. Lim et al., *Process Control for Sheet-Metal Stamping*,
Advances in Industrial Control, DOI: 10.1007/978-1-4471-6284-1_8,

adapted) on-line without intermediate computations involving process parameter estimates. In Chap. 7 automatic tuning (auto-tuning) and direct AC of stamping were addressed (Lim et al. 2012). The auto-tuning was used not only to reduce the effort in tuning the controller gains, but also to determine good initial values of the controller gains for use with a direct AC. As shown in Fig. 8.1a, the direct AC, or model reference adaptive control (MRAC), was used to adaptively update the controller gains (which are initially set to the values obtained from auto-tuning) to minimize the tracking error between the reference model output (y_m) and the process output (y), in the presence of disturbances. The direct MRAC scheme has been shown to meet the performance requirements, which include stability and asymptotic tracking, for minimum-phase systems. However, in general, a parameterization for process parameter estimation in the adaptive law is not possible for nonminimum-phase systems with direct MRAC (Astrom and Wittenmark 1995; Ioannou and Sun 1996). In addition, since the regressor vector in the adaptation algorithm for the direct MRAC uses the nominal values of the process parameters, the robustness and sensitivity analysis of the direct MRAC to process model perturbations, which represent unexpected plant dynamics changes and/or a slowly time-varying system, has been considered in (Lim et al. 2012).

On the other hand, indirect AC, which provides estimates of the process parameters and thus does not require the nominal process parameters, is applicable to both minimum- and nonminimum-phase systems. The indirect AC uses process observation with on-line estimates of process parameters, and then it is necessary to introduce safeguards to make sure that all conditions required for the controller design method are fulfilled (Astrom and Wittenmark 1995). For example, it may be necessary to test whether the estimated process model is minimum-phase or whether there are common factors in the estimated polynomials. In addition, with indirect AC, solutions for the adaptation mapping between the estimated process parameters ($\theta(t)$) and the controller parameter ($\theta_c(t)$), defined by an algebraic equation $\theta_c(t) = f\{\theta(t)\}$, cannot be guaranteed to exist at each time (t), thus, giving rise to two potential issues: *loss of stabilizability* and *non-uniqueness*.

Many studies have considered the stabilizability issue using various ideas. One possibility is to modify the adaptation algorithm so that the parameter estimates are projected into a given fixed constrained region (Goodwin and Sin 1984; Kreisselmeier 1985; Chia et al. 1991). For example, it may be sufficient to project into a set such that $\theta_{min} \leq \theta(t) \leq \theta_{\max}$. Another idea is to add a so-called leakage term to the adaptation law to keep the estimates near a priori estimates. The idea behind leakage is to modify that adaptive law so that the time derivative of the Lyapunov function, which is used to analyze the adaptive scheme, becomes negative in the space of parameter estimates when these parameters exceed certain bounds (Ioannou and Kokotovic 1984). A third method is to introduce a dead-zone in the estimator, switching off the parameter estimation if the error is too large (Astrom and Wittenmark 1995; Ioannou and Sun 1996). However, the ideas introduced above require prior knowledge to select constraints and bounds, and require satisfying a persistent excitation criterion.

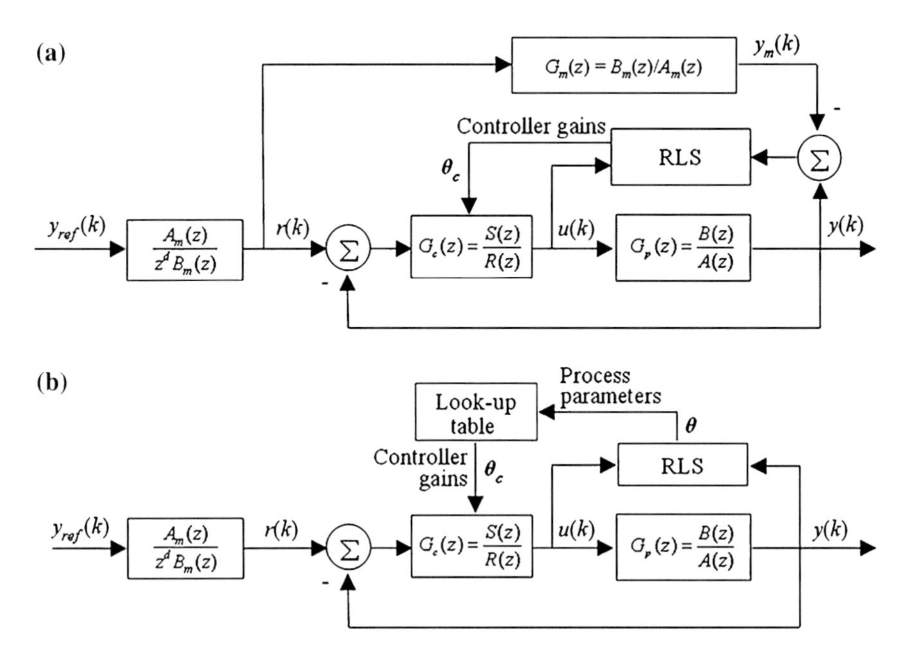

Fig. 8.1 Design methods for adaptive process controllers: **a** direct model reference adaptive control (MRAC), **b** indirect method using look-up table

The non-uniqueness problem arises where a unique solution of the algebraic equation in the adaptation law does not exist, regardless of prior knowledge. To guarantee existence and uniqueness of solutions to the algebraic mapping equation requires various assumptions in the process model $\left(\text{i.e.,} G_p(z) = B(z)/A(z)\right)$ and the controller structure $\left(\text{i.e.,} G_c(z) = S(z)/R(z)\right)$ (Astrom and Wittenmark 1995; Ioannou and Sun 1996). For example, the polynomials A and B are required to be coprime, with the assumption that A is monic. In addition, the degrees of the polynomials R and S in the controller must be constrained with respect to the degrees of the process polynomials A and B. Consequently, if R and S are selected for a certain control structure, without consideration of such constraints on the degrees of R and S, then the adaptation law for adjustment of the controller gain from the process parameter may require an optimization procedure (Astrom and Wittenmark 1995).

In this section, indirect AC, which provides estimates of the process parameters and is applicable to both minimum- and nonminimum-phase systems, is addressed. For our application, a proportional plus integral (PI) controller, which has been successfully used for rejecting disturbances and improving robustness to model uncertainty in stamping (Sunseri et al. 1996; Hsu et al. 2000, 2002; Lim et al. 2010), is selected. Furthermore, in a recent study Lim (2010), Lim et al. (2012) utilized the auto-tuning method to tune the gains of a standard PI controller. Such auto-tuning can also be utilized to provide good initial values of the PI gains for a direct adaptive controller. However, the simple PI control structure selected for our indirect adaptive controller requires an optimization approach for the computation of the controller gains, and is thus not amenable to on-line implementation. Hence, an indirect AC using a look-up table scheme, which updates the controller gains on-line based on the process parameter estimates, is, for the first time, introduced and evaluated. Noting that simple control structures (e.g., P, PI and PID) are used in many control applications, the proposed look-up table-based indirect AC described in following sections should be of wide interest.

As shown in Fig. 8.1b, the look-up table is used to obtain the controller gains, $\theta_c(t)$, in terms of the estimated parameters, $\theta(t)$. This is done on-line, as part of the indirect AC which estimates the process parameters $\theta(t)$. The look-up table itself is created off-line using optimization to solve the adaptation mapping, $\theta_c(t) = f\{\theta(t)\}$, which has a non-unique solution due to the selection of a simple PI control structure for our application. Then, this functional relationship is discretized in the look-up table via a number of break points within a given constrained (or projected) range based on prior knowledge obtained from experiments. Subsequently, this indirect AC, with various levels of process parameter discretization in the look-up table, is investigated in simulations. However, due to extensive memory requirements in real-time implementation, a smaller look-up table is used in the experiments. This look-up table-based indirect AC is compared with direct MRAC in terms of tracking performance as well as part quality, in the presence of plant variations (e.g., changes in lubrication and material thickness). In addition, in order to achieve high output tracking performance (Tomizuka 1987; Devasia 2002; Karimi et al. 2008), a pre-compensator based on the inverse of the closed-loop system is also included.

8.2 Prior Information for Adaptive Control

Many adaptive control schemes (both direct and indirect) require a priori information about the process dynamics and utilize a pre-determined controller structure. In our previous work (Lim et al. 2010, 2012), system identification based on standard least squares (LS) was used to parameterize the plant dynamics of a MIMO linear sheet metal stamping using input (i.e., binder force) and output (i.e., punch force) data from experiments. Furthermore, the simulation-based autotuning, also referred to as the pre-tune mode, was introduced to obtain the controller gains, which were also used as appropriate initial values for the design of the adaptation law in the direct AC (i.e., MRAC). Moreover, disturbances (e.g., lubrication and thickness change), which affect not only the process variable (i.e., punch force) but also part quality were addressed. The key features of that work are briefly summarized in this section.

8.2.1 Perturbation Process Model Structure

The perturbation process model that relates the change in commanded binder force, $\delta F_{b,ref}$ (input, u), to the change in filtered punch force, $\delta F_{p,fil}$ (output, y), is given by (Lim et al. 2009, 2010, 2012):

$$\frac{y(k)}{u(k)} = \frac{\delta F_{p,fil}(k)}{\delta F_{b,ref}(k)} = \frac{B(z)}{A(z)} = \frac{b_{11}z^2 + b_{10}z}{z^4 + a_3 z^3 + a_2 z^2 + a_1 z + a_0} \tag{8.1}$$

This perturbation process model includes the machine control (i.e., inner-loop hydraulic actuator machine control) and a low-pass noise filter as used in the experiments.

Next, the model in Eq. (8.1) is extended to the MIMO case by creating a 4×12 transfer function matrix (TFM), with four punch forces $\left(\delta F^i_{p,fil} i = 1,2,3,4\right)$ as outputs and 12 binder forces $\left(\delta F^i_{bj,ref} j = 1,2,3\right)$ as inputs. Based on the experimentally verified assumption that each punch force output is primarily affected by the three nearest binder forces as inputs, the TFM is constrained to a block-diagonal form given by

$$\begin{bmatrix} \delta F^1_{p,fil}(k) \\ \delta F^2_{p,fil}(k) \\ \delta F^3_{p,fil}(k) \\ \delta F^4_{p,fil}(k) \end{bmatrix} = \begin{bmatrix} G^1_1 & G^1_2 & G^1_3 & 0 & 0 & 0 & 0 & 0 & 0 & 0 & 0 & 0 \\ 0 & 0 & 0 & G^2_1 & G^2_2 & G^2_3 & 0 & 0 & 0 & 0 & 0 & 0 \\ 0 & 0 & 0 & 0 & 0 & 0 & G^3_1 & G^3_2 & G^3_3 & 0 & 0 & 0 \\ 0 & 0 & 0 & 0 & 0 & 0 & 0 & 0 & 0 & G^4_1 & G^4_2 & G^4_3 \end{bmatrix} \begin{bmatrix} \delta F^1_{b1,ref}(k) \\ \cdot \\ \cdot \\ \delta F^4_{b3,ref}(k) \end{bmatrix} \tag{8.2}$$

Thus, the MISO estimated linear process model in discrete-time [decoupled MIMO transfer function matrix in Eq. (8.2)] for each corner output or y^i (i = 1, 2, 3, 4) is given as

$$y^i(k) = \frac{B^i_1(z)u^i_1(k)}{A^i(z)} + \frac{B^i_2(z)u^i_2(k)}{A^i(z)} + \frac{B^i_3(z)u^i_3(k)}{A^i(z)} \tag{8.3}$$

where

$$\begin{aligned} B^i_j(z) &= b^i_{j1}z^2 + b^i_{j0}z \, j = 1,2,3 \\ A^i(z) &= z^4 + a^i_3 z^3 + a^i_2 z^2 + a^i_1 z + a^i_0 \\ y^i(k) &= \delta F^i_{p,fil}(k) \\ \left[u^i_1(k)\, u^i_2(k)\, u^i_3(k)\right]^T &= \left[{}^i_{b1,ref}(k)\, \delta F^i_{b2,ref}(k)\, \delta F^i_{b3,ref}(k)\right]^T \end{aligned}$$

The unknown parameters in the 4th order perturbation process models shown in Eq. (8.3) are parameterized based on experimental data using the standard Least Squares (LS) algorithm (Lim et al. 2010, 2012). Each estimated model characterizes the dynamics of the process from the change of three reference binder force inputs to the change of one filtered punch force output. The estimated parameters of the 4th order perturbation process models in discrete-time have been experimentally validated by matching the experimentally measured punch force outputs with the punch force generated by the estimated models using the same commanded binder forces in the desired case.

8.2.2 Process Controller Structure

As shown in Fig. 8.1, when a pre-compensator for tracking is considered, providing the reference input (r^i), the proportional plus integral (PI) controller structure, based on three binder force inputs (u_j^i) for each corner punch force output (y^i), is given by

$$u_j^i(k) = C_j^i(z)\left\{r^i(k) - y^i(k)\right\} \quad i = 1, 2, 3, 4 \tag{8.4}$$

where

$$C_j^i(z) = \frac{S(z)}{R(z)} = \left(\frac{s_{j1}^i z + s_{j0}^i}{z - 1}\right) = \left(K_P^i + \frac{K_I^i}{z - 1}\right) \quad j = 1, 2, 3$$

Note that the SIMO PI controller parameters are related to the traditional proportional (K_P) plus integral (K_I) gains for each corner i as:

$$s_{j1}^i = K_{Pj}^i \text{ and } s_{j0}^i = K_{Ij}^i - K_{Pj}^i, \quad j = 1, 2, 3 \tag{8.5}$$

8.3 Direct and Indirect Adaptive Control

In this section, the design and implementation of direct and indirect AC to improve part quality in stamping, in the presence of disturbances, is compared. First, the design and implementation of a direct MRAC process controller, which was described in the Chap. 7, is briefly summarized. Second, the design and implementation of an indirect AC using a look-up table is presented.

8.3.1 Direct Model Reference Adaptive Control (MRAC)

As shown in Fig. 8.1a, the direct MRAC process controller, which updates its controller gains to make the measured punch force (y) track the reference model output (y_m) as closely as possible in the presence of plant dynamics variations and disturbances, is described in the Chap. 7. In the direct MRAC, a reference model specifies the desired performance of the closed-loop system. Thus, the reference model has the same structure as the closed-loop system with a fixed-gain PI controller, and is obtained by combining Eqs. (8.3) and (8.4), with the values of s_{j1}^i and s_{j0}^i set to their values obtained from auto-tuning, and by using the nominal process model parameters. The reference model based on the gains chosen using auto-tuning satisfies the step response specifications required for the stamping process, with a settling time of less than 0.1 s and an overshoot of less than 20 %. The most critical phase of forming is the latter phase, well after the transients from the initial impact of the press have settled. The entire forming process in a typical production press takes between 0.5–1 s, and thus a 20 % overshoot for 0.1 s is acceptable and has been validated in experimental tests.

8.3.2 Indirect Adaptive Control Using Look-Up Table

On-Line Plant Parameter Estimation

For the indirect AC shown in Fig. 8.1b, process parameters are estimated on-line based on the observation of the input (u) and output (y) in experiments. Based on the unknown parameters and measurements for each corner output i, the regression model of the MISO plant transfer function shown in Eq. (8.3) is given by

$$\begin{aligned} y^i(k) = & -a_3^i y^i(k-1) - a_2^i y^i(k-2) - a_1^i y^i(k-3) - a_0^i y^i(k-4) \\ & + \left\{ b_{11}^i u_1^i(k-2) + b_{10}^i u_1^i(k-3) + \ldots + b_{31}^i u_3^i(k-2) + b_{30}^i u_3^i(k-3) \right\} \end{aligned} \tag{8.6}$$

Using the regression model in Eq. (8.6), the parametric model for process parameter estimation is formulated in terms of estimated parameters $\left(\text{i.e.}, \hat{a}_3^i, \ldots, \hat{a}_0^i, \hat{b}_{11}^i, \ldots, \hat{b}_{30}^i\right)$ as

$$y^i = \theta^{iT} \phi^i \tag{8.7}$$

where

$$\begin{aligned} \theta^{iT} &= \left[\hat{a}_3^i, \ldots, \hat{a}_0^i, \hat{b}_{11}^i, \ldots, \hat{b}_{30}^i\right] \\ \phi^i &= \left[-y^i(k-1), \ldots, -y^i(k-4), u_1^i(k-2), \ldots, u_3^i(k-3)\right]^T \end{aligned}$$

Recursive computation for indirect AC, based on the parametric model above, uses the RLS algorithm with exponential forgetting. The process parameters, θ^i, are estimated recursively using the estimation error, $\varepsilon^i = y^i - \hat{y}^i$:

$$\begin{aligned} \theta^i(k) &= \theta^i(k-1) + P^i(k)\phi^i(k) \left\{ y^i(k) - \underbrace{\left(\phi^{iT}(k)\theta^i(k-1)\right)}_{\hat{y}^i} \right\} \\ P^i(k) &= \left[P^i(k-1) - \frac{P^i(k-1)\phi^i(k)\phi^{iT}(k)P^i(k-1)}{\lambda I + \phi^{iT}(k)P^i(k-1)\phi^i(k)} \right] \Big/ \lambda \end{aligned} \tag{8.8}$$

Adaptation Law Using Pole Placement

Referring to Fig. 8.1b, $A^i \in R^1, B^i_j \in R^{1\times 3}$ and $C^i_j \in R^{3\times 1}$ are defined in Eqs. (8.3) and (8.4) for each corner i. Thus, the closed-loop transfer function in discrete-time is given by

$$\frac{y^i(k)}{r^i(k)} = \frac{B^i_1(z)C^i_1(z) + B^i_2(z)C^i_2(z) + B^i_3(z)C^i_3(z)}{A^i(z) + B^i_1(z)C^i_1(z) + B^i_2(z)C^i_2(z) + B^i_3(z)C^i_3(z)} \tag{8.9}$$

Hence, the closed-loop characteristic polynomial is:

$$A^i + B^i_1 C^i_1 + B^i_2 C^i_2 + B^i_3 C^i_3 = A^i_c \tag{8.10}$$

Substituting A^i, B^i_j and C^i_j given in Eqs. (8.3) and (8.4) respectively into Eq. (8.10), the closed-loop characteristic polynomial, based on on-line estimated parameters in Eq. (8.7) and the controller gains in Eq. (8.4), becomes

$$A^i_c(z) = z^5 + a^i_{c4} z^4 + a^i_{c3} z^3 + a^i_{c2} z^2 + a^i_{c1} z + a^i_{c0} \tag{8.11}$$

where

$$\begin{aligned}
a^i_{c4} &= \hat{a}^i_3 - 1 \\
a^i_{c3} &= \hat{a}^i_2 - \hat{a}^i_3 + \hat{b}^i_{11} s^i_{11} + \hat{b}^i_{21} s^i_{21} + \hat{b}^i_{31} s^i_{31} \\
a^i_{c2} &= \hat{a}^i_1 - \hat{a}^i_2 + \hat{b}^i_{11} s^i_{10} + \hat{b}^i_{21} s^i_{20} + \hat{b}^i_{31} s^i_{30} + \hat{b}^i_{10} s^i_{11} + \hat{b}^i_{20} s^i_{21} + \hat{b}^i_{30} s^i_{31} \\
a^i_{c1} &= \hat{a}^i_0 - \hat{a}^i_1 + \hat{b}^i_{10} s^i_{10} + \hat{b}^i_{20} s^i_{20} + \hat{b}^i_{30} s^i_{30} \\
a^i_{c0} &= -\hat{a}^i_0
\end{aligned}$$

Assuming that the desired closed-loop poles are placed at five locations $\left(\text{i.e.}, p^i_d, d = 1, \ldots, 5\right)$ in the z-plane, the desired closed-loop characteristic polynomial is given by

$$\begin{aligned}
A^i_d(z) &= \left(z - p^i_1\right)\left(z - p^i_2\right)\left(z - p^i_3\right)\left(z - p^i_4\right)\left(z - p^i_5\right) \\
&= z^5 + l^i_4 z^4 + l^i_3 z^3 + l^i_2 z^2 + l^i_1 z + l^i_0
\end{aligned} \tag{8.12}$$

where all of the desired poles in discrete-time lie inside the unit circle. The five desired poles are grouped as two dominant poles $\left(\text{i.e.}, \left|p^i_{1,2}\right| \to 1 \text{ in } z - \text{plane}\right)$ and three fast (or non-dominant) poles $\left(\text{i.e.}, \left|p^i_{3,4,5}\right| \to 0 \text{ in } z - \text{plane}\right)$. The two dominant poles are selected based on specifications for the stamping process, i.e., a settling time of less than 0.1 s and an overshoot of less than 20 %.

The controller gains $\left(\text{i.e.}, s^i_{j1} \text{ and } s^i_{j0}, j = 1, 2, 3\right)$ are determined by matching coefficients of the two polynomials in Eqs. (8.11) and (8.12) respectively. Thus,

$$
\begin{aligned}
&z^4: -(p_1^i+p_2^i+p_3^i+p_4^i+p_5^i) = \hat{a}_3^i - 1\\
&z^3: (p_1^ip_2^i+p_2^ip_3^i+p_3^ip_4^i+\ldots) = \hat{a}_2^i-\hat{a}_3^i+\hat{b}_{11}^is_{11}^i+\hat{b}_{21}^is_{21}^i+\hat{b}_{31}^is_{31}^i\\
&z^2: -(p_1^ip_2^ip_3^i+p_1^ip_2^ip_4^i+p_1^ip_2^ip_5^i+\ldots) = \hat{a}_1^i-\hat{a}_2^i+\hat{b}_{11}^is_{10}^i+\hat{b}_{21}^is_{20}^i+\hat{b}_{31}^is_{30}^i+\hat{b}_{10}^is_{11}^i+\hat{b}_{20}^is_{21}^i+\hat{b}_{30}^is_{31}^i\\
&z^1: (p_1^ip_2^ip_3^ip_4^i+p_1^ip_2^ip_3^ip_5^i+\ldots) = \hat{a}_0^i-\hat{a}_1^i+\hat{b}_{10}^is_{10}^i+\hat{b}_{20}^is_{20}^i+\hat{b}_{30}^is_{30}^i\\
&z^0: -(p_1^ip_2^ip_3^ip_4^ip_5^i)
\end{aligned}
\tag{8.13}
$$

In Eq. (8.13), there are six unknown controller gains $\left(\text{i.e.}, s_{j1}^i \text{ and } s_{j0}^i, j=1,2,3\right)$, but only five linear equations.

In general, such linear systems can be solved using a generalized inverse (or pseudo-inverse). Note that Eq. (8.13) can be expressed as $Ax = b$, where x is the vector of unknown controller gains $\left(\text{i.e.}, \left[s_{11}^i,\ldots,s_{30}^i\right]^T\right)$ and A is a coefficient matrix. However, the 5×6 matrix A is not full rank (i.e., rank $(A) = 3$) because there are no unknowns $(\text{i.e.}, s_{11}^i,\ldots,s_{30}^i)$ in the z^4 and z^0 coefficients in Eq. (8.13). Thus, determining the controller gains, in terms of the estimated process parameters and the desired poles, requires an optimization procedure as follows (Quintana-Ortí et al. 1998).

Optimization Approach

The error for each term shown in Eq. (8.13) is defined as

$$
\begin{aligned}
\gamma_4^i &= \hat{a}_3^i - 1 + (p_1^i+p_2^i+p_3^i+p_4^i+p_5^i)\\
\gamma_3^i &= \hat{a}_2^i-\hat{a}_3^i+\hat{b}_{11}^is_{11}^i+\hat{b}_{21}^is_{21}^i+\hat{b}_{31}^is_{31}^i-(p_1^ip_2^i+p_2^ip_3^i+p_3^ip_4^i+\ldots)\\
\gamma_2^i &= \hat{a}_1^i-\hat{a}_2^i+\hat{b}_{11}^is_{10}^i+\hat{b}_{21}^is_{20}^i+\hat{b}_{31}^is_{30}^i+\hat{b}_{10}^is_{11}^i+\hat{b}_{20}^is_{21}^i+\hat{b}_{30}^is_{31}^i+(p_1^ip_2^ip_3^i+p_1^ip_2^ip_4^i+p_1^ip_2^ip_5^i+\ldots)\\
\gamma_1^i &= \hat{a}_0^i-\hat{a}_1^i+\hat{b}_{10}^is_{10}^i+\hat{b}_{20}^is_{20}^i+\hat{b}_{30}^is_{30}^i-(p_1^ip_2^ip_3^ip_4^i+p_1^ip_2^ip_3^ip_5^i+\ldots)\\
\gamma_0^i &= -\hat{a}_0^i + (p_1^ip_2^ip_3^ip_4^ip_5^i)
\end{aligned}
\tag{8.14}
$$

Then, based on both the desired poles $\left(\text{i.e.}, p_d^i, d=1,2,3,4,5\right)$ and the estimated plant parameters $\left(\text{i.e.}, \hat{A}^i \text{ and } \hat{B}_j^i, j=1,2,3\right)$, the sum of the squares of the five errors (i.e., $\gamma_k^i = f\left(\hat{a}_3^i,\ldots,\hat{b}_{30}^i,p_1^i,\ldots,p_5^i\right), k = 0,1,2,3,4$) can be minimized to find the controller gains $\left(\text{i.e.}, s_{j1}^i \text{ and } s_{j0}^i, j=1,2,3\right)$ for each corner output i, i.e.,

$$
\min_{s_{j1}^i, s_{01}^i}\left\{\left(\gamma_0^i\right)^2+\left(\gamma_1^i\right)^2+\left(\gamma_2^i\right)^2+\left(\gamma_3^i\right)^2+\left(\gamma_4^i\right)^2\right\} \tag{8.15}
$$

Thus, the computation of the controller gains requires an optimization approach, which is not amenable to on-line implementation. Consequently, the adaptation mechanism using a look-up table scheme, which updates on-line the controller gains on-line based on the process parameter estimates is developed and evaluated.

Adaptation Mechanism Using Look-Up Table

As shown in Fig. 8.1b, the controller gains, which are calculated off-line via optimization and are stored in a look-up table, can be selected on-line based on the values of the estimated plant parameters. The look-up table is generated via optimization and implemented as follows:

Step 1: Generate a mesh size (i.e., discretization) for the process parameters. The process parameters need to have a certain number of discrete values within appropriate bounds based on prior knowledge. For example, the process parameters $\left(\text{i.e.}, \hat{a}_3^i, \ldots, \hat{b}_{30}^i\right)$ are bounded within a certain variation (i.e., $\pm 2\delta$ or $\pm 20\,\%$) where δ is the variation of each discretized parameter, using n (e.g., $n = 3$ or 5) discretized points around the nominal values $\left(\text{i.e.}, \bar{a}_3^i, \ldots, \bar{b}_{30}^i\right)$ of the parameters. Such discretized sets of bounded process parameters used to generate the look-up tables are shown in Table 8.1. Subsequently, the closed-loop performance of the look-up table will be investigated and compared in terms of the number of discrete break points n.

Step 2: Choose the desired pole locations for optimization. To obtain solutions for the controller gains $\left(\text{i.e.}, s_{j1}^i \text{ and } s_{j0}^i\right)$ via optimization (see Eq. 8.15), the five desired poles in the z-plane must be specified. Two dominant poles (i.e., $0.68 \pm 0.2i$) and three fast, or non-dominant, poles (i.e., 0.1, 0.3 and 0.4) are chosen in the z-plane. These satisfy system specifications required for the stamping process, i.e., a settling time of less than 0.1 s and an overshoot of less than 20 %.

Step 3: Perform the optimization so that the look-up table stores the controller gains $\left(\text{i.e.}, s_{j1}^i \text{ and } s_{j0}^i\right)$ based on both the discretized sets of process parameters and the given desired poles. This was accomplished using the standard function *fminsearch* in MATLAB for unconstrained optimization (Lagarias et al. 1998): This function starts at given initial values and then finds local minima of the given error equations. Thus, prior information for the controller gains, obtained using auto-tuning, is used for initial values for the optimization.

Step 4: Formulate the look-up table based on the discretized values of the estimated process parameters. This was accomplished using the standard Simulink block (i.e., *look-up table (n-D)* in MATLAB). As shown in Fig. 8.2, the standard look-up table block generates an output value $\left(s_{11}^i\right)$ by comparing the block inputs $\left(\hat{a}_3^i, \hat{a}_2^i, \ldots, \hat{b}_{31}^i, \hat{b}_{30}^i\right)$ with the discretized set parameters. The look-up table evaluates a sampled representation of a function in m variables (i.e., the estimated process parameters or $m = 10$ here) by linear interpolation between samples to give an approximate value. For example, one of the controller gains can be approximated by

$$s_{11}^i = f(x_1, x_1, \ldots, x_{m-1}, x_m) = f\left(\hat{a}_3^i, \hat{a}_2^i, \ldots, \hat{b}_{31}^i, \hat{b}_{30}^i\right) \tag{8.16}$$

where m is the number of estimated parameters.

The data parameter in the look-up table is defined as a set of output values that correspond to its rows, column, and higher dimensions (or pages) with the nth

Table 8.1 Discretized sets of the bounded process parameters used in the look-up table

Plant parameter	$n = 3$	$n = 5$	...
$\hat{a}_3^i$	$\begin{bmatrix} \hat{a}_3^i(1) \\ \hat{a}_3^i(2) \\ \hat{a}_3^i(3) \end{bmatrix} = \begin{bmatrix} \bar{a}_3^i - 2\delta \\ \bar{a}_3^i \\ \bar{a}_3^i + 2\delta \end{bmatrix}$	$\begin{bmatrix} \hat{a}_3^i(1) \\ \hat{a}_3^i(2) \\ \hat{a}_3^i(3) \\ \hat{a}_3^i(4) \\ \hat{a}_3^i(5) \end{bmatrix} = \begin{bmatrix} \bar{a}_3^i - 2\delta \\ \bar{a}_3^i - \delta \\ \bar{a}_3^i \\ \bar{a}_3^i + \delta \\ \bar{a}_3^i + 2\delta \end{bmatrix}$	...
...	...	...	...
$\hat{b}_{30}^i$	$\begin{bmatrix} \hat{b}_{30}^i(1) \\ \hat{b}_{30}^i(2) \\ \hat{b}_{30}^i(3) \end{bmatrix} = \begin{bmatrix} \bar{b}_{30}^i - 2\delta \\ \bar{b}_{30}^i \\ \bar{b}_{30}^i + 2\delta \end{bmatrix}$	$\begin{bmatrix} \hat{b}_{30}^i(1) \\ \hat{b}_{30}^i(2) \\ \hat{b}_{30}^i(3) \\ \hat{b}_{30}^i(4) \\ \hat{b}_{30}^i(5) \end{bmatrix} = \begin{bmatrix} \bar{b}_{30}^i - 2\delta \\ \bar{b}_{30}^i - \delta \\ \bar{b}_{30}^i \\ \bar{b}_{30}^i + \delta \\ \bar{b}_{30}^i + 2\delta \end{bmatrix}$	...

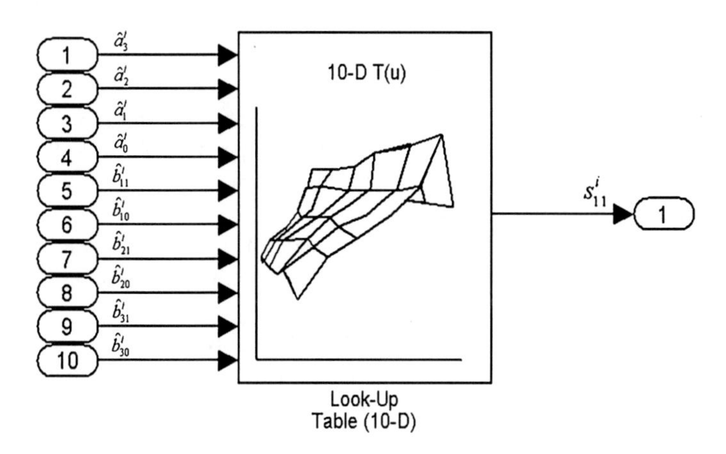

Fig. 8.2 An example of a set of input and output values for the look-up table

discretized set parameter shown in Table 8.1. As shown in Table 8.2, the first $(\hat{a}_3^i)$ input identifies the first dimension (row) break points, the second $(\hat{a}_2^i)$ input identifies the second dimension (column) break points, the third input $(\hat{a}_1^i)$ identifies the third dimension (or page) break points, and so on (see Simulink block *look-up table* in MATLAB for more detail). Table 8.2 shows, for two different discretization values n (i.e., $n = 3$ and 5), the values of s_{11}^i (one of the six controller gains for each corner i) based on the 10 estimated process parameters. In Table 8.2, $\hat{a}_3^i$ (first) and $\hat{a}_2^i$ (second) inputs specify row and column discretized points respectively, and the eight other inputs specify their first value $\left(\text{i.e., } \hat{a}_1^i(1), \hat{a}_0^i(1), \ldots, \hat{b}_{31}^i(1), \hat{b}_{30}^i(1)\right)$ of discretized points. Due to space limitations all values cannot be shown. However, all of the different values for Table 8.2

Table 8.2 An example of the controller gain (i.e., s_{11}^{i}) stored in the look-up table based on two different discretizations (i.e., $n = 3$ and 5) for variations of two plant parameters

		S_{11}^{i}			
	Break points (n = 3)	Column ($\widehat{a}_2^l$)	$\widehat{a}_2^l(1)$	$\widehat{a}_2^l(2)$	$\widehat{a}_2^l(3)$
n = 3	Row($\widehat{a}_3^l$)		$\overline{a}_2^l - 2\delta$	$\overline{a}_2^l$	$\overline{a}_2^l + 2\delta$
	$\widehat{a}_3^l(1)$	$\overline{a}_3^l - 2\delta$	0.152	0.166	0.178
	$\widehat{a}_3^l(2)$	$\overline{a}_3^l$	0.152	0.153	0.166
	$\widehat{a}_3^l(3)$	$\overline{a}_3^l + 2\delta$	0.172	0.152	0.153
		…			

	Break points (n = 5)	Column ($\widehat{a}_2^l$)	$\widehat{a}_2^l(1)$	$\widehat{a}_2^l(2)$	$\widehat{a}_2^l(3)$	$\widehat{a}_2^l(4)$	$\widehat{a}_2^l(5)$
n = 5	Row($\widehat{a}_3^l$)		$\overline{a}_2^l - 2\delta$	$\overline{a}_2^l - \delta$	$\overline{a}_2^l$	$\overline{a}_2^l + \delta$	$\overline{a}_2^l + 2\delta$
	$\widehat{a}_3^l(1)$	$\overline{a}_3^l - 2\delta$	0.152	0.153	0.166	0.171	0.178
	$\widehat{a}_3^l(2)$	$\overline{a}_3^l - \delta$	0.151	0.152	0.158	0.153	0.172
	$\widehat{a}_3^l(3)$	$\overline{a}_3^l$	0.152	0.151	0.153	0.152	0.166
	$\widehat{a}_3^l(4)$	$\overline{a}_3^l + \delta$	0.162	0.155	0.152	0.151	0.158
	$\widehat{a}_3^l(5)$	$\overline{a}_3^l + 2\delta$	0.172	0.161	0.152	0.155	0.153
		…					

are calculated via optimization based on the first, second or third values of discretized points for the eight other inputs. Thus, the data parameters in the look-up tables require a large memory size, which will be discussed in following section.

Simulation with Look-Up Table

Simulation is used to validate the performance of the look-up table scheme in Fig. 8.1b, based on the estimated perturbation process models, with randomly-assigned process model parameter variations to represent changes in plant dynamics. The simulations are performed in terms of three cases: (1) Case A is the original indirect AC based on optimization without using a look-up table. This case is not for the real-time implementation in experiments, but for validation of look-up table performance. (2) Case B is for a look-up table with $n = 3$. (3) Case C is for a look-up table $n = 5$. Linear interpolation in the look-up tables is used to pick an appropriate controller gain between values of the controller gains based on the estimated process parameters. Case A (i.e., off-line optimization) uses the original values of the PI controller gains based on the estimated plant parameters, without any interpolation.

As shown in Fig. 8.3, the look-up table using $n = 5$ (Case C) is closer to Case A than $n = 3$ (Case B) in the look-up table. In particular, Fig. 8.3a, b and e show the expected results that the look-up table with $n = 5$ is closer to the values from the optimization than the look-up table with $n = 3$. In addition, as shown in Fig. 8.4, the tracking performance of the punch force with $n = 5$ in the look-up table is slightly better than the look-up table with $n = 3$.

However, the case with five discretized points requires extensive memory. Specifically, for each corner i, the number of data points required is $l \times n^2 \times n^m$, where l is the number of the controller gains (or the number of look-up tables), n is

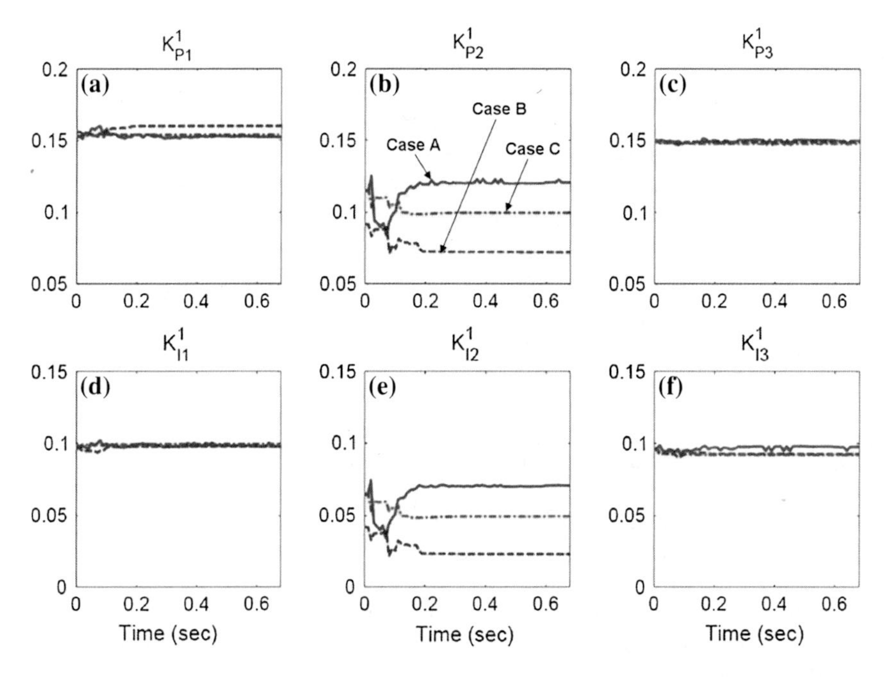

Fig. 8.3 Simulated results of the controller gains based on three cases (i.e., (1) *Case A* (solid) with off-line optimization (2) *Case B* (dashed) with look-up table and $n = 3$ (3) *Case C*: (dash-dot) with look-up table and $n = 5$)

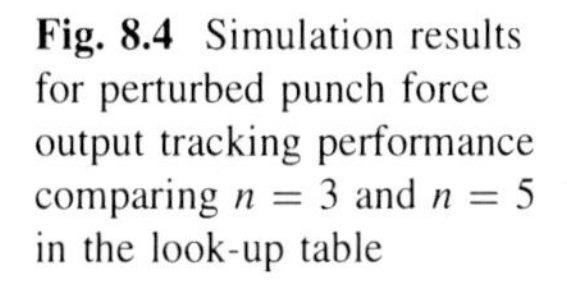

Fig. 8.4 Simulation results for perturbed punch force output tracking performance comparing $n = 3$ and $n = 5$ in the look-up table

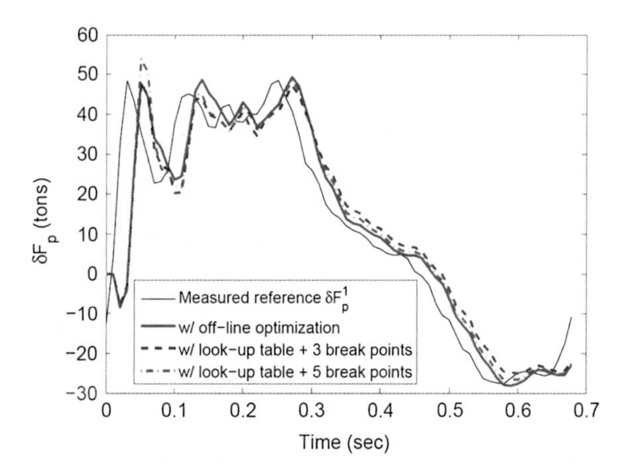

the number of discretization points, and m is the number of plant parameters. For $n = 5$ this becomes 1.5×10^9, while with $n = 3$ the memory requirement is for 3.2×10^6 points, for $l = 6$, and $m = 10$. However, due to memory limitations in the real-time control computer, even a look-up table with three discretized points could not be accommodated in experiments. Consequently, a smaller look-up table (i.e., $n = 2$) was used in the experiments (memory requirement is for 2.5×10^4 points). It is compared with the direct AC in terms of the tracking performance as well as part quality, in the presence of disturbances, in a subsequent section.

In this simulation the estimated process parameters using system identification are used as initial values of the process parameters in the RLS algorithm, and estimated process models are used as the process model. Initial values of the covariance matrix P^i (see Eq. 8.8) are set to $10^3 \times I\,(m,m)$ where I is the identity matrix and m (i.e., $m = 10$) is the number of the process parameters for each corner. In simulations, normally distributed random numbers (with a variance less than 1 % of output (y)), are added to the input (u) to satisfy persistent excitation conditions.

8.4 Simulation and Experimental Results

8.4.1 Simulation Results

Simulation is used to validate and compare the performance of the two proposed adaptive PI process controllers (i.e., direct and indirect) based on the estimated perturbation process models, with randomly-assigned process model parameter variations to represent changes in plant dynamics. The simulation models use the perturbed binder forces $\left(\delta F^i_{bj}\right)$ as inputs, the perturbed punch force $\left(\delta F^i_p\right)$ as output, and the desired perturbed punch force $\left(\delta F^i_{p,ref}\right)$ as the reference (or desired punch force) for each corner i.

The three binder forces $\left(F^i_{bj}, j = 1, 2, 3\right)$ associated with each punch force corner, i, ($i = 1, 2, 3, 4$) are updated to minimize the difference between F^i_p and $F^i_{p,ref}$, by adding or subtracting the perturbed binder forces $\left(\delta F^i_{bj}\right)$, which are produced by the process controller. Thus, the total binder forces, F^i_{bj}, in both simulation and experiment are given by

$$F^i_{bj} = F^i_{bj,offset} + \delta F^i_{bj} j = 1, 2, 3 \tag{8.17}$$

where $F^i_{bj,offset}$ are pre-determined nominal binder forces for each corner, i, and are shown in Fig. 8.5. As described previously, the machine control (MC) alone (without the process controller) generates these pre-determined nominal binder forces, even in the presence of disturbances.

The three binder forces $\left(F^i_{bj}, j = 1, 2, 3\right)$ associated with each punch force corner, i, ($i = 1, 2, 3, 4$) are updated to minimize the difference between F^i_p and $F^i_{p,ref}$, by adding or subtracting the perturbed binder forces $\left(\delta F^i_{bj}\right)$, which are

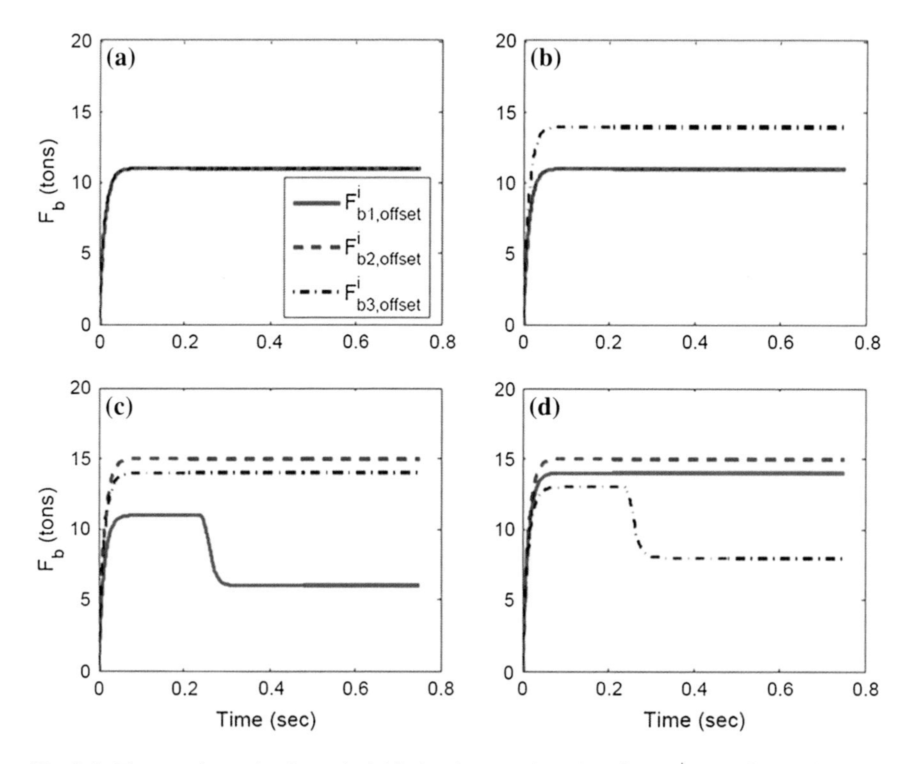

Fig. 8.5 The pre-determined nominal binder force trajectories (i.e., $F^i_{bj,offset}$) for each corner output i: **a** $i = 1$, **b** $i = 2$, **c** $i = 3$, **d** $i = 4$

produced by the process controller. Thus, the total binder forces, F^i_{bj}, in both simulation and experiment are given by

Note that the reference punch force trajectories $\left(F^i_{p,ref}\right)$ are obtained by recording the punch forces generated using the pre-determined nominal binder forces which are determined by experienced die-makers to make a good part using the material with nominal properties and under normal operating conditions for each punch force corner i.

Figures 8.6 and 8.7 show the simulation results with the punch force as output and the binder force as input, comparing different tracking performance between the direct AC (i.e., MRAC) and the look-up table-based indirect AC (i.e., $n = 3$) for minimum- and nonminimum-phase system models as shown in Table 8.3. With the system identification technique described in our previous studies (Lim et al. 2009, 2010, 2012), the minimum- and nonminimum-phase model were obtained with different sample rates—100 Hz and 1 kHz respectively. Furthermore, these simulations include 20 % variation in each model as shown in Table 8.3. In Fig. 8.6, with the minimum-phase system model, both process controllers track the punch force by adjusting the binder forces. However, Fig. 8.7a shows that the look-up table-based indirect AC (i.e., $n = 3$) controller outperforms the direct AC (i.e.,

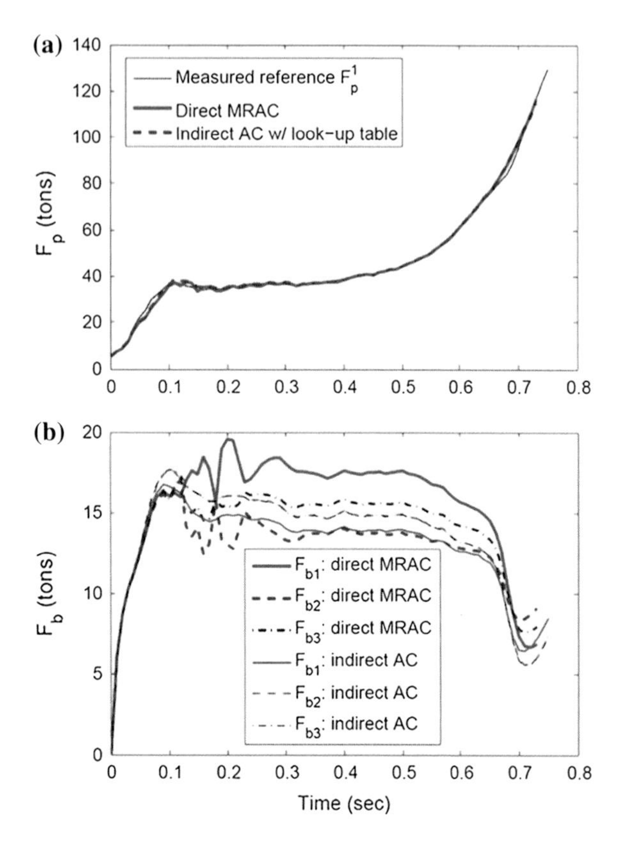

Fig. 8.6 Simulation results comparing two process controllers (i.e., direct and indirect AC with $n = 3$ in the look-up table) with *minimum-phase* system model as shown Table 8.3: **a** punch force (i.e., output or F_p^1), **b** binder forces associated with F_p^1

MRAC) for the nonminimum-phase system model. In Fig. 8.7b, as in the experimental system, binder forces are constrained using a saturation block (i.e., maximum = 20 tons) in the Simulink model to ensure that no mechanical damage is caused by over-driving the actuators. Due to space limitations results are presented here for only one corner of the punch forces (i.e., F_p^1). However, all corners of the punch forces were investigated and showed similar trends. The performance of the two controllers in the presence of actual plant variations and disturbances in the form of intentionally introduced lubrication and material thickness changes is described in Sect. 8.4.2 on experimental validation.

8.4.2 Experimental Validation

Direct and indirect AC process controllers described above were implemented on the experimental system. Their performance was compared to the performance of the machine control (MC) only [i.e., without process control (PC)) with fixed predetermined binder forces commands ($F^i_{bj,offset}$ in Eq. (8.17)], in terms of deviation

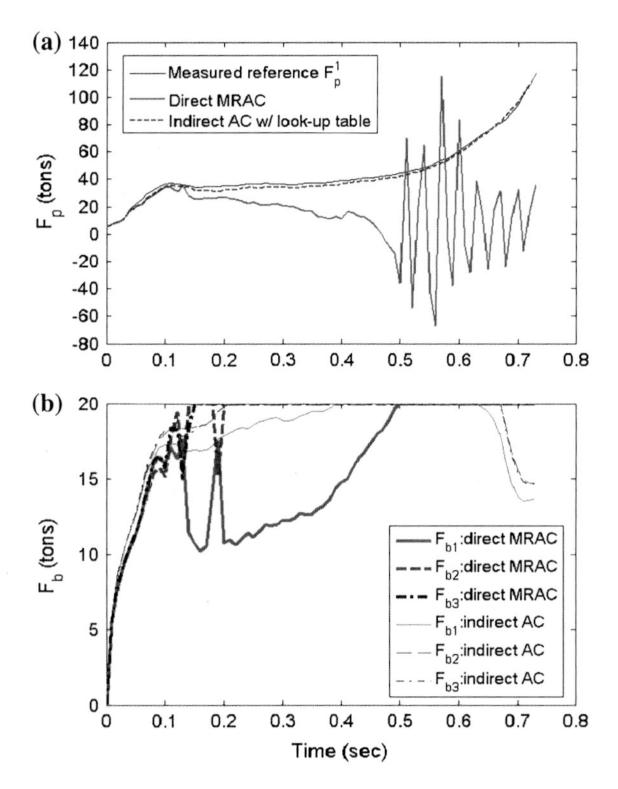

Fig. 8.7 Simulation results comparing two process controllers (i.e., direct and indirect AC with $n = 3$ in the look-up table) with *nonminimum-phase* system model as shown Table 8.3: **a** punch force (i.e., output or F_p^1), **b** binder forces associated with F_p^1

from the reference punch force, and in terms of part quality, in the presence of disturbances.

Lubrication Change

The first disturbance that we consider is lubrication change. Figure 8.8 shows the tracking performance of the punch force for the test cases, and illustrates stamped part quality comparison for those cases in the presence of lubrication change. Due to space limitations results are presented for only one corner of the punch forces (i.e., F_p^1). However, all corners of the punch forces and part quality for all corners were investigated and showed similar trends.

Figure 8.8a shows that with only machine control, (i.e., without process control) there is a clear difference between the reference punch force which characterizes a good part in the absence of the extra lubrication and the measured punch force in the presence of the extra lubrication. Figure 8.8e shows that extra lubrication results in significant wrinkling, caused by greater material flow. The desired part quality, characterized by the reference punch force, is shown in Fig. 8.8b. In Fig. 8.8a and c it can be seen that the direct MRAC process controller provides the best tracking performance of the reference trajectory as well as significant part quality improvement in the presence of lubrication change. The look-up table-based indirect AC process controller with $n = 2$ also enables the punch force to

Table 8.3 Estimated process model parameters (i.e., $i = 1$) with different sample rate (Ts) as shown in Eq. (8.3) for both minimum- and nonminimum-phase model

Parameter	b_{11}^{1}	b_{10}^{1}	b_{21}^{1}	b_{20}^{1}	b_{31}^{1}	b_{30}^{1}	a_{3}^{1}	a_{2}^{1}	a_{1}^{1}	a_{0}^{1}
Minimum-phase ($Ts = 100$ Hz)	0.7301	−0.7262	0.4272	−0.5516	0.8058	−0.8294	−0.3315	−0.4481	−0.3071	0.2897
Nonminimum-phase ($Ts = 1{,}000$ Hz)	0.9255	−0.9562	0.9567	−0.9116	0.9234	−0.9594	−0.4524	−0.5424	−0.6445	0.7456

Fig. 8.8 Experimental results in the presence of lubrication change: **a** the punch force (i.e., F_p^1), and for indirect AC with $n = 2$, **b** dry condition and **c**–**e** excessive lubricated condition

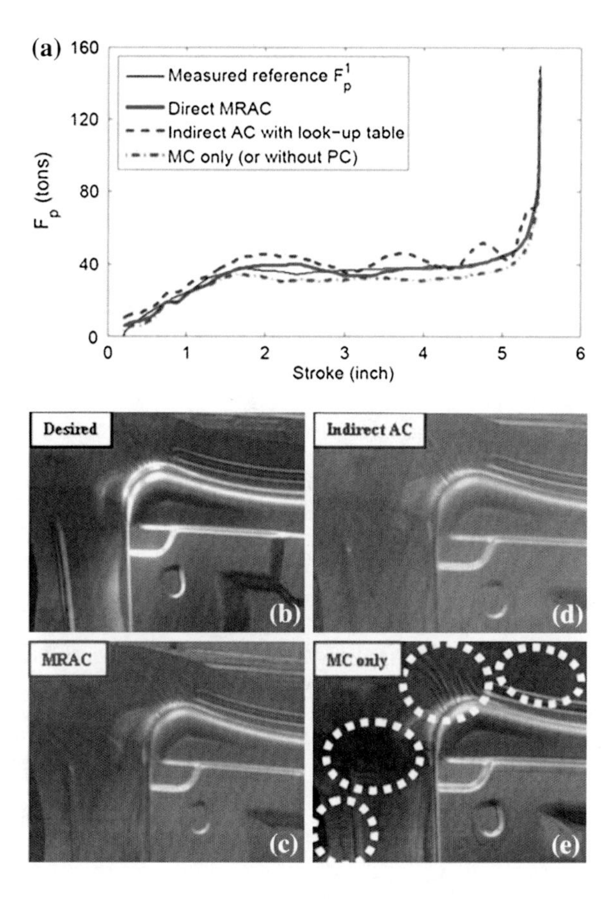

track the reference punch force, but with some oscillation throughout the punch stroke under the excessive lubrication condition. Note that part quality is improved with both AC methods over the MC only case. Thus, the above experimental results, using a complex geometry part, show that the direct MRAC process controller outperforms the look-up table-based indirect AC with $n = 2$ in terms of the tracking of the reference punch force trajectory, while both process controllers effectively improve part quality in the presence of lubrication change.

Material Thickness Variation

The second disturbance that we consider is material thickness variation. With thicker material (i.e., 0.79 mm) compared to the nominal (i.e., 0.64 mm), Fig. 8.9 shows that the two process controllers effectively track the reference punch force, which was determined using the nominal material. Again, as shown in Fig. 8.9, machine control alone (i.e., without PC) cannot minimize the error between the reference punch force and the measured punch force. However, the direct MRAC process controller also shows good tracking performance of the measured

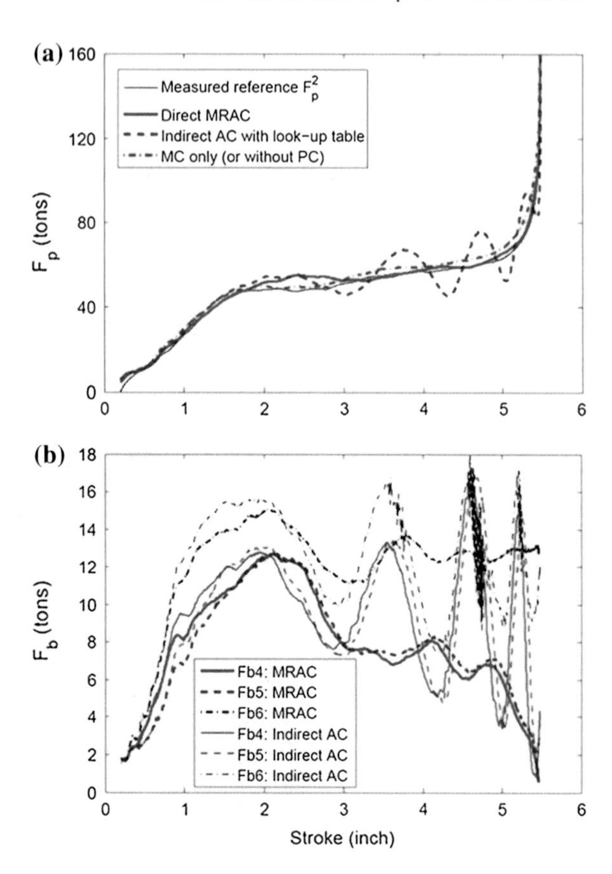

Fig. 8.9 Experimental results in the presence of material thickness variation: **a** punch force (i.e., F_p^2), and for indirect AC with $n = 2$, **b** binder forces

reference punch force by adjusting the three associated binder forces (see Fig. 8.9b). For example, when the measured punch force is greater than the reference punch force, at around 2.5 in. of punch stroke, the direct MRAC process controller drastically reduces the binder forces in order to minimize the difference between the reference F_p^2 and measured F_p^2. However, the indirect AC process controller based on a look-up table with $n = 2$ performs less effectively with material thickness change, and shows excessive oscillations. Due to memory limitations, the indirect AC with a smaller look-up table ($n = 2$) is used in the experiments, where it is outperformed by the direct AC.

The performance of the look-up table-based indirect AC is restricted, due to memory limitations of the real-time control computer. Required memory sizes of the look-up table are proportional to not only the number of discretizations for the plant parameters, but also the number of plant parameters as well as the number of the controller gains. However, in this paper the results presented, for the first time, show a novel methodology where indirect AC requires an optimization approach to obtain the solution (i.e., the controller parameter) to the adaptation mapping equation, due to certain constraints (e.g., constrained structures in the controller and/or the process model). Clearly, the adaptive controller parameters generated via optimization, which is not amenable to real-time implementation, can be embedded into the indirect AC in the form of a look-up table based on the estimated parameters, and implemented in real-time production runs. Such an approach can be expected to become more attractive in the future, as memory becomes cheaper and faster.

8.5 Discussion and Remarks

Initially, to study the look-up table method, we considered a 1st order SISO process model having two estimated parameters combined with a SISO PI control structure. The SISO PI process controller gains obtained via off-line optimization were tuned manually and then stored in the look-up table, with three or five discretized points for the two process parameters. In other words, every possible case for the discretized points of two process parameters was considered one-by-one to obtain the PI gains via the optimization. However, for our stamping application, it was not practical to tune the controller gains manually to generate the look-up table for each corner. Consequently, the process used to generate the look-up tables of the controller gains via off-line optimization was automated. This automation, which computes every case for the discretized points (e.g., $n = 3$ or 5) of the 10 estimated parameters, and generates SIMO PI controller gains matrix for the look-up table, makes the look-up table scheme feasible, even with large number of discretized points for the process parameters.

In addition, simulations with indirect AC using the look-up table method show it works well. However, the size of the look-up table becomes a critical issue for real-time implementation. Thus, the performance in experiments, which will be shown in Sect. 8.6, is not as good as desired due to the limit of $n = 2$. Ultimately, indirect AC with the look-up table may be more attractive in the future as memory for real-time computer control continues to become cheaper and faster.

A nonminimum-phase system could result in this stamping application due to oversampling (i.e., sampling at too high a frequency). It may also occur in other applications, not only due to oversampling, but also due to inherent dynamic characteristics of the system being controlled. The process dynamics are dependent on several factors that can vary from part to part, for example, die geometry, material characteristics and press dynamics. The nominal process models for each case are obtained using system identification and there is no guaranty that the model will be minimum-phase. Thus, indirect AC, which is not restricted to minimum-phase systems, can be valuable in practice.

The computational burden for the proposed Indirect AC occurs in generating the look-up table, which is done off-line and leads to large on-line memory requirements. As noted in the discussion on simulation and experimental results, the use of higher-dimensional tables obviously results in better performance of the Indirect AC; however, the number of elements in the look-up tables is given by

$l \times n^2 \times n^m$, where l is the number of the controller gains (or the number of look-up tables), n is the number of discretization points, and m is the number of plant parameters, and thus, the memory requirements grow rapidly as n increases. Accessing the stored gains from the look-up table in real-time is done very quickly (e.g., a few micro seconds) and does not constitute a significant computational burden. In fact, look-up tables can be implemented on a Field Programmable Gate Array (FPGA) with very low execution times; however, the biggest constraint currently is the memory required for the size of the look-up table.

8.6 Summary and Conclusions

A comparison between direct and indirect adaptive control (AC) in stamping in terms of the design, implementation, tracking performance and improvement in part quality is presented. Both simulation and experimental results in the presence of disturbances and process variations are included. A MIMO direct AC (i.e., MRAC) stamping process controller works well. However, the adaptation algorithm for the direct AC requires the nominal process parameters. Consequently, an indirect AC, which does not need the nominal process parameters, is considered as a potential alternative. However, depending on the control structure, indirect AC may require one to find controller gains using an optimization approach, which is difficult to do in a real-time implementation. Thus, the indirect AC using a look-up table scheme is proposed.

In this chapter, we have described, for the first time, the design and implementation of a look-up table-based indirect AC, which updates the controller gains on-line using a look-up table generated via off-line optimization. In simulation, the indirect AC, with a sufficiently high level of discretization (i.e., $n = 3$ or 5) for the estimated process parameters in the look-up table, performs well for both minimum- and nonminimum-phase system model while the direct MRAC is restricted to nonminimum-phase system model. However, the size of the look-up table becomes an issue for real-time implementation. Thus, performance of indirect AC, with a lower level of discretization (i.e., $n = 2$) for the estimated process parameters in a look-up table, is not as good as the direct AC. In the future, indirect AC with the look-up table may become more attractive as memory for real-time computer control continues to become cheaper and faster.

References

Astrom, K. J., Wittenmark, B., 1995, *Adaptive Control*, Prentice Hall, 2nd edition.

Boling, J.M., Seborg, D.E., Hespanha, J.P., 2007, "Multi-model adaptive control of a simulated pH neutralization process," *Control Engineering Practice*, Vol. 15, Issue 6, Special Section on Control Applications in Marine Systems—CAMS2004, Control Applications in Marine Systems, June 2007, pp. 663–672.

Chia, T.L., Chow, P.-C., Chizeck, H.J., 1991, "Recursive Parameter Identification of Constrained Systems: An Application to Electrically Stimulated Muscle," *IEEE Trans. Biomed. Eng.*, Vol. 38, pp. 429–442.

Devasia, S., 2002, "Should Model-Based Inverse Inputs Be Used as Feedforward Under Plant Uncertainty?," *IEEE Trans. on Automatic Control*, Vol. 47, no. 11, pp. 1865–1871.

Goodwin, G.C., Sin, K.W., 1984, *Adaptive Filtering, Prediction, and Control*, Prentice-Hall: Englewood Cliffs, NJ.

Hsu, C.W., Ulsoy, A.G., Demeri, M.Y., 2000, "An Approach for Modeling Sheet Metal Forming for Process Controller Design," *ASME J. Manuf. Sci. Eng.*, Vol. 122, pp. 717–724.

Hsu, C. W., Ulsoy, A. G., Demeri, M. Y., 2002, "Development of process control in sheet metal forming," *J. of Materials Proc. Tech.*, Vol. 127, pp. 361–368.

Ioannou, P.A., Kokotovic, P.V., 1984, "Instability Analysis and Improvement of Robustness of Adaptive Control," *Automatica*, Vol. 20, no. 5, pp. 583–594.

Ioannou, P.A., Sun, J., 1996, *Robust Adaptive Control*, New Jersey: Prentice-Hall.

Karimi, A., Butcher, M., Longchamp, R., 2008, "Model-Free Pre-compensator Tuning Based on the Correlation Approach," *IEEE Trans. on Control Systems Tech.*, Vol. 16, pp. 1013–1020.

Kreisselmeier, G., 1985, "An Approach to Stable Indirect Adaptive Control," *Automatica*, Vol. 21, no. 4, pp. 425–431.

Lagarias, J.C., Reeds, J.A., Wright, M.H., Wright, P.E., 1998, "Convergence Properties of the Nelder-Mead Simplex Method in Low Dimensions," *SIAM Journal of Optimization*, Vol. 9, no. 1, pp. 112–147.

Lauderbaugh, L.K., Ulsoy, A.G., 1989, "Model Reference Adaptive Force Control in Milling," *ASME J. of Engineering for Industry*, Vol. 111, no. 1, pp. 13–21.

Lim, Y.S., 2010, *MIMO Adaptive Process Control in Stamping Using Punch Force*, Ph.D. Dissertation, University of Michigan, Ann Arbor, Michigan.

Lim, Y.S., Venugopal, R., Ulsoy, A.G., 2009, "Improved Part Quality in Stamping Using Multi-Input Multi-Output Process Control," *Proceedings of American Control Conference*, Saint Louis, MO, USA, June 10-12, 2009, pp. 5570–5575.

Lim, Y.S., Venugopal, R., Ulsoy, A.G., 2010, "Multi-Input Multi-Output Modeling and Control for Stamping," *ASME J. of Dyn. Systems, Meas., and Control*, Vol. 132, Issue 4, 041004.

Lim, Y.S., Venugopal, R., Ulsoy, A.G., 2012, "Auto-Tuning and Adaptive Stamping Process Control," *Control Engineering Practice*, Vol. 20, pp. 156–164.

Quintana-Ortí, G., Quintana-Ortí, E. S., Petitet, A., 1998, "Efficient Solution of The Rank-Deficient Linear Squares Problem," *SIAM J. Sci. Comput.*, Vol. 20, no. 3, pp. 1155–1163.

Sunseri, M., Cao, J., Karafillis, A.P., Boyce, M.C., 1996, "Accommodation of Springback Error in Channel Forming Using Active Binder Force: Control Numerical Simulations and Experiments," *J. of Engin. Materials and Tech.*, Vol. 118, pp. 426–435.

Rupp, D. and Guzzella, L., 2010, "Adaptive internal model control with application to fueling control," *Control Engineering Practice*, Vol. 18, Issue 8, pp. 873–881.

Tomizuka, M., 1987, "Zero Phase Error Tracking Algorithm for Digital Control," *ASME J. of Dynamic Systems, Measurement, and Control*, Vol. 109, pp. 65–68.

Tsai, Ching-Chih, Lin, Shui-Chun, Wang, Tai-Yu, Teng, Fei-Jen, 2009, "Stochastic model reference predictive temperature control with integral action for an industrial oil-cooling process," *Control Engineering Practice*, Vol. 17, Issue 2, pp. 302–310.

Chapter 9
Concluding Remarks

Abstract This book has introduced process control for sheet metal stamping, including theory and experiments. Real-time, in-process adjustment of blank holder forces enables one to achieve the desired flow of the blank material into the die without tearing and wrinkling even in the presence of process disturbances, such as changes in lubrication and/or material properties (e.g., material formability, sheet thickness). Such a process control system for stamping can be achieved using a reconfigurable array of hydraulic actuators to provide variable binder force capability together with punch force sensors to provide in-process feedback. Multi-input multi-output process controllers (fixed gain or adaptive) can be used to ensure consistent operation, even in hard to form materials, despite process disturbances. Such a system can be retrofitted for use with existing mechanical presses and can also be implemented in modern hydraulic presses. Benefits include elimination of tearing, wrinkling and springback with reduced die try-out times, as well as consistent part quality in production even in the presence of disturbances. The potential economic benefits of such systems can be enormous. The controller designs presented herein have been shown to be effective in experimental tests on production automotive stamped parts. However, in this chapter we also discuss some potential limitations and future areas of research and development that will further enhance this technology.

9.1 Summary

Sheet metal stamping, due to low-cost and high-productivity, is a widely used and economically important process for manufacturing automotive body panels, white goods, and many other consumer products (Kalpakjian and Schmid 2001; Hosford and Caddel 2011). To produce high-quality parts (i.e., without tearing, wrinkling and springback) the dies must be carefully designed (e.g., using finite element methods) and refined and vetted in a die-try-out process, especially for difficult-to-form lightweight alloys. This book has described how control technologies can be

Y. Lim et al., *Process Control for Sheet-Metal Stamping*,
Advances in Industrial Control, DOI: 10.1007/978-1-4471-6284-1_9,

used to adjust the flow of the blank material into the die to reduce die-try-out times and to consistently achieve high-part quality even in the presence of process variations.

The system presented is reconfigurable, consisting of individually controlled hydraulic actuators, which replace traditional nitrogen cylinders and can be positioned to support the die at various locations around the binder. The number of actuators required (e.g., 10–20) depends on the complexity of the part being manufactured. Each actuator is individually controlled by a servo-valve to achieve the desired binder force at that location. Furthermore, the control is high-bandwidth, such that the binder force can be varied as a function of time (or stroke) during the short duration (e.g., 0.5–1 s) of the stamping process. Higher binder forces make it more difficult for material to flow into the die at that location, while reduced binder forces at a particular location make material flow into the die cavity easier. Varying the binder forces during the punch stoke can help set the part geometry and reduce springback.

Commanding and controlling these binder forces, at various locations and as function of time (or stoke), is what we have termed *machine control* and described in detail in Chap. 4. Experienced operators can quickly learn how to use these machine control capabilities to eliminate tearing, wrinkling and springback and to reduce the die-try-out process (without welding and grinding of dies) from several days to a few hours.

In a production environment, there can be changes in lubrication or in the properties (e.g., formability or sheet thickness) of the sheet metal blanks. Such in-process disturbances can lead to high scrap rates in production. However, by adding in-process feedback, through measurement of punch forces at the four corners of the press, one can automatically adjust the binder forces to reject the effects of process disturbances. This is termed *process control* and is described in detail in Chap. 6. The process controller adjusts the binder forces as needed to maintain, despite disturbances, the desired punch force trajectories for producing a good part. Techniques such as auto-tuning and adaptive control can be used to minimize the test parts needed for designing the gains of the controller.

9.2 Benefits of Process Control in Stamping

Process control in stamping complements current techniques such as finite element analysis for die design, die-try-out, and press monitoring. The installation costs of a stamping process controller are a small percentage (e.g., 5 %) of the cost of the press. The process control software can be readily learned by, and empowers, press operators. Due to techniques such as auto tuning and adaptive control, it does not require extensive engineering time to setup and maintain the system. The reconfigurable hydraulic actuators used to control blank holder forces can replace nitrogen cylinders ("nitro cushions") and be readily setup for different dies and part geometries. The process control methods presented here can be used in existing mechanical presses as well as in new hydraulic presses.

The benefits of using process control in stamping with variable binder force are numerous. Specifically:

1. Improved part quality and reduced scrap rates
2. Use of smaller/thinner blanks for material savings
3. Enabling stamping of hard-to-form materials, e.g., lightweight alloys
4. Reductions in die-try-out times and costs
5. Reductions in finite element analysis simulation times
6. Increase of press life by reducing press impact forces.

The exact economic benefits, and return on investment, of using process control in stamping will depend on the specific product and production environment. However, for a given part an estimate can be made as in the following example of a production automotive panel. The part is assumed to be a door-inner made of aluminum and is manufactured under the following conditions:

1. Tandem-line press runs at 10 strokes/min, or production of 600 parts/h
2. Differential between cost and scrap recovery rate per blank is $35
3. Scrap rate without process control is 8 %.

From test data, it has been shown that the use of variable blank holder force yields material savings from blank size reduction of 2 % and process control can reduce scrap from 8 to 4 %. Thus, total material savings of at least 6 % are achieved or in dollar terms = 0.06 × $35/part × 600 parts/h = $1,080/h.

Other benefits include reduced try-out time for launching dies, reduced down-time for die work, lower press maintenance costs, and lower costs for production delays. With a utilization of 1,000 h/year, the use of process control with variable binder force in production generates savings of more than $1 million per year per stamping line. While these numbers are based on specific cases, significant benefits can be expected in most situations. Further information on the economic benefits can be found on the web site for Intellicass Inc. (Venugopal 2013).

One significant benefit of the technology, which is difficult to value monetarily, is that it improves safety for shop-floor personnel. The traditional approach to correcting part defects by "working" the die, that is, grinding and welding, requires an experienced tool-maker to work in the die, increasing the risk of both traumatic and chronic injuries. Anecdotal evidence from tool-makers on the technology points to widespread approval, as it makes their job easier.

Shop-floor experience with the systems has shown that tool-makers can be very quickly trained to use process control with variable binder-force actuation. With appropriate user interfaces as described in Chap. 4, an experienced die-maker can run the system with a day of training.

Process control for stamping has been proven to bring valuable benefits to the manufacturing industry while presenting several paths of research that can lead to further improvements.

9.3 Future Research and Extensions

During the course of the experimental tests described in previous chapters, the authors interacted with several types of technical personnel involved in the production of stamped parts, and these interactions revealed that there are several challenges in moving this technology to widespread use on the production floor.

The first challenge involves integrating variable binder force technology into the die design process. The current paradigm is to ensure fixed and even binder force, with material flow control effected through the use of a drawbead. The drawbead is designed using finite-element analysis (FEA) simulations. Extensive simulation time is necessary to accurately design a drawbead, and while this method has become precise enough to produce high quality parts at production launch with minimal try-out, a fixed drawbead cannot compensate for defects that occur due to operational variability. Process control will address operational variations but its effect can be maximized by designing the die to incorporate variable binder force. First, FEA can be used to ensure that the die is sufficiently flexible to maximize the effect of variable binder force in problem areas. Next, initial variable binder force trajectories can be estimated using FEA (Ahmetoglu et al. 1995; Sunseri et al. 1996; Krishnan and Cao 2003; Palaniswamy et al. 2004), reducing both try-out and die design time, as part quality issues arising from assumptions during design can be rapidly corrected during try-out using variable binder force. Including variable binder force at the die design stage also allows for reducing material costs through the use of smaller blanks and thinner gauges (as long as other considerations such as safety requirements are met). In addition, more relaxed tolerances can be utilized in the specification of the sheet metal, further reducing material costs. Finally, data collected from the digital process control system can be used to improve FEA model fidelity.

The next area of improvement can come in embedding sensors in the die for more accurate defect prediction and better process control feedback. In-die piezo-based load cells that measure the friction force between the sheet metal and the die surface have been described in (Siegert et al. 1997), while optical sensors have been described in (Doege et al. 2003). A comprehensive study of in-die sensors for monitoring forming dynamics and part quality can be found in Mahayotsunan et al. (2009). By placing sensors in the die near problem areas, more localized process control can be implemented, thus reducing the risk of correcting a defect in one area while creating another in another area, as might happen while using a more global measurement from tonnage monitors. In-die feedback sensors for process control have not been tested yet in actual production environments and more research is required into developing robust, cost-effective sensors and optimizing their location in the die.

One of the limitations of the hydraulic-cylinder based design for the actuation system is that involves connecting hoses between the cylinders in the die and a hydraulic manifold near the bolster of the press. These connections, even with quick-disconnects, can take a few extra minutes during die change in production.

Designing a complete cost-effective in-die hydraulic actuation system would make it much easier to deploy process control technology in existing presses in stamping plants, as retrofitting the presses with an additional hydraulic system will not be necessary. Creating such a system may entail the development of cost-effective servo-valves that can be built into the hydraulic cylinders and a minimally sized hydraulic tank that can be pressurized with shop-air. Thermal issues would need to be addressed as the thermal effects on the hydraulic system will be more pronounced as the volume of fluid used is reduced.

On the algorithmic side, as noted in Chaps. 7 and 8, the minimum-phase assumption for the direct MRAC approach cannot be verified a priori. Thus, developing a robust control scheme that can handle the change in plant parameters due to operational variability in the stamping process is definitely an area of further research. The indirect adaptive control scheme described in Chap. 8 seeks to address the minimum-phase limitation; however, as stated, the computational requirements for implementing a sufficiently fine look-up table in real-time are currently a limitation.

The process control approaches described in this book require system identification techniques and future work can include using FEA simulation to obtain the required parameters prior to try-out, eliminating the need for experimental tests for the purpose of system identification. Such an approach would be more shop-floor friendly.

The process control design approach described in this book also lends itself to extension to other related manufacturing processes such as hydro-forming or warm-forming. In the case of hydroforming, variable blank-holder force has been shown to improve part quality. In warm-forming, the distribution of thermal energy across the die may be adjusted to compensate for operational variations and maintain part quality.

References

Ahmetoglu, M., Broek, T. R., Kinzel, G., Altan, T. (1995), Control of blank holder force to eliminate wrinkling and fracture in deep-drawing rectangular parts. CIRP Annals-Manufacturing Technology, 44(1): 247–250

Doege, E., Schmidt-Jürgensen, R., Huinink, S.,Yun, J. W. (2003) Development of an optical sensor for the measurement of the material flow in deep drawing processes. CIRP Annals-Manufacturing Technology, 52(1): 225–228.

Hosford, W. F., Caddell, R. (2011) Metal forming: mechanics and metallurgy. Cambridge University Press, Cambridge.

Kalpakjian, S., Schmid, S.R. (2001) Manufacturing engineering and technology. Prentice-Hall, New Jersey

Krishnan, N. and Cao, J. (2003) Estimation of optimal blank holder force trajectories in segmented binders using an ARMA model, ASME Journal of Manufacturing Science and Engineering 125(4): 763–771

Mahayotsanun, N., Sah, S., Cao, J., Peshkin, M., Gao, R. X., Wang, C. T. (2009). Tooling-integrated sensing systems for stamping process monitoring. International Journal of Machine Tools and Manufacture, 49(7): 634–644.

Palaniswamy, H., Ngaile, G., Altan, T. (2004), Finite element simulation of magnesium alloy sheet forming at elevated temperatures. Journal of Materials Processing Technology, 146(1): 52–60.

Siegert, K., Ziegler, M., Wagner, S. (1997) Closed loop control of the friction force. Deep drawing process. Journal of materials processing technology, 71(1):126–133.

Sunseri, M., Cao, J, Karafillis, A.P. Boyce, M. (1996), Accommodation of springback error in channel forming using active binder force control: Numerical simulation and experiments, ASME Journal of Engineering Materials and Technology 118(1):426–435.

Venugopal, R. (2013) Intellicass web site, http://www.intellicass.com/

Index

Y. Lim et al., *Process Control for Sheet-Metal Stamping*,
Advances in Industrial Control, DOI: 10.1007/978-1-4471-6284-1,